Technical Artefacts: Creations of Mind and Matter

Philosophy of Engineering and Technology

VOLUME 6

For further volumes:
http://www.springer.com/series/8657

Peter Kroes

Technical Artefacts: Creations of Mind and Matter

A Philosophy of Engineering Design

Springer

Peter Kroes
Department of Philosophy
Delft University of Technology
Delft, The Netherlands

ISSN 1879-7202 ISSN 1879-7210 (electronic)
ISBN 978-94-007-3939-0 ISBN 978-94-007-3940-6 (eBook)
DOI 10.1007/978-94-007-3940-6
Springer Dordrecht Heidelberg New York London

Library of Congress Control Number: 2012939387

Printed on acid-free paper

Springer is part of Springer Science+Business Media (www.springer.com)

Dedicated to:

Sven
Isabel
Hannah

Preface

The view on the nature of technical artefacts developed in the following pages is a "creation of the mind" that I have been working on for over more than a decade. Elements of this view have been published previously but here an attempt is made to present them as parts of a coherent vision on what kind of objects technical artefacts are, how they come into being, and how they are related to physical and social objects. The dual nature view I propose may be seen as a reaction to, on the one hand, accounts of technical artefacts as social constructions, which have become very popular in particular in Science and Technology Studies (STS), and, on the other hand, conceptions of (technical) artefacts as creations of the mind, which are rather dominant in philosophical quarters. Both accounts of technical artefacts suffer from a common bias for the intentional features of technical artefacts at the expense of their material features. In my opinion any convincing view on technical artefacts and how they come into being must take due account of the role of material features. According to the dual nature account both kinds of features are constitutive for being a technical artefact.

This book has grown out of a research project entitled *The Dual Nature of Technical Artefacts* that was funded by the Dutch National Science Foundation (NWO) and started in the year 2000. During a five year period a group of philosophers worked together on various problems related to technical artefacts. I am immensely indebted to each and every member of this group: Anthonie Meijers, Wybo Houkes, Pieter Vermaas, Jeroen de Ridder, Marcel Scheele and Maarten Franssen. I have had the privilege, together with Anthony, of leading this group. Being one of its members has been a great learning experience. I would not have been able to write this book without this 'Dual-Nature group' whose members shared a common view on how to approach problems in the philosophy of technology. I am very grateful to all of them for the many discussions we have had on philosophical issues about the nature of technical artefacts and on philosophical issues in general. Likewise I am very grateful to Joseph Pitt, Davis Baird, Carl Mitcham, Louis Bucciarelli and Randall Dipert, our American colleagues, who from the very beginning have been associated with the Dual-Nature project and whose ideas, comments and criticism have always been a source of inspiration.

The Dual-Nature project was executed at the department of Philosophy of Delft University of Technology in the Netherlands. This is a great and exciting place to do work in the philosophy of technology. Its members share an interest in and focus on philosophical problems about technology and I am grateful to all of them for numerous fruitful discussions. Furthermore I am indebted to Pieter Vermaas as editor of this book series and to Springer for their help and support in turning the manuscript into this printed text. A special word of thanks goes to Maarten Franssen for his incisive comments on the final version of the whole manuscript. Over the years I have come to appreciate his philosophical knowledge and expertise. I would also like to thank Jeroen van den Hoven. His readiness to chair the department of philosophy made it possible for me to take a sabbatical to work on this book. Finally, I thank the Netherlands Institute for Advanced Study (NIAS) for making it possible for me to write a first version of this book during my stay as a fellow in residence in the period 2006-2007.

Peter Kroes

Contents

Chapter 1
Introduction

The world in which we live is a world of technical artefacts. We live our lives with and through them. It is not so much the natural world as well as the technical world that conditions human life. This technical world not only provides us with the means to adapt the physical environment to our needs and desires. Its influence stretches out much further into the world of social affairs and into the world of ideas. Through the ages, for instance, technology has provided strong metaphors for interpreting what it means to be a human being, such as the man-machine or the brain-computer metaphors. So, the technical world strongly influences human thinking and doing. This book is an attempt to understand what kind of world this technical world is by studying the nature of the basic elements that make up this world, namely technical artefacts. Given their pervasive influence on human thoughts and actions, such an understanding may contribute to, or may even be a requisite step to a better understanding of the modern human condition. To that end, this book addresses a number of questions all of which centre around technical artefacts. What kind of objects are they? What does it mean for an object to be a technical artefact? In what sense are they different from objects from the natural world, or the social world? How do they come into existence? Does it make sense to consider technical artefacts to be morally good or bad because of the way they influence human life?

The common way to set technical artefacts apart from natural objects is a genetic one, by taking into account their history. This is already signified by the term 'artefact' itself, meaning literally 'made by craft/skill'. More in particular, technical artefacts are considered to be things made by humans for supporting them in fulfilling their practical needs and ends, that is, objects made by humans for practical reasons (purposes). They differ from natural things that may also be used for solving practical problems by the fact that they are made by human beings. On this view, the difference between technical artefacts and natural things concerns the way they come into being. Technical artefacts come into being through human activity by being invented and made by humans. Natural things, on the contrary, come into being without human intervention; it would not make any sense to claim that they are our inventions. This way of characterizing the difference between technical

P. Kroes, *Technical Artefacts: Creations of Mind and Matter: A Philosophy of Engineering Design*, Philosophy of Engineering and Technology 6,
DOI 10.1007/978-94-007-3940-6_1, © Springer Science+Business Media B.V. 2012

and natural objects runs closely parallel to the standard way of distinguishing technology from science. A scientist primarily positions him/herself as a spectator, a discoverer or a theorist in this world, as someone who has cognitive needs and desires and who only actively intervenes in the world in so far this is necessary for observing things and phenomena and for creating things and phenomena that could have occurred in nature but actually are not occurring.[1] A technician or engineer, on the contrary, is always out to change the (material) world in order to adapt that world to the practical needs and desires of humans. He or she is an actor in this world, not a spectator, and changes what happens on the scene, or better, changes the scene itself by continually creating new kinds of technical artefacts.[2]

So the technical world is a world produced by humans. It is part of the broader world of human making, the artificial world, which is populated by artefacts in general.[3] This artificial world contains all kinds of subclasses of artefacts, two of which are worth mentioning here because they are more or less close 'cousins' of technical artefacts, but which nevertheless fall outside the scope of this book, namely works of art and social artefacts. The artificial world also includes works of art. A general characterization of works of art is not an easy matter. A traditional way of demarcating works of art from technical artefacts is by characterizing works of art as objects made for aesthetic reasons whereas technical artefacts are objects made for practical reasons. These reasons, however, do not exclude each other, which may make it difficult to classify human-made objects unambiguously as works of art or as technical artefacts. Our focus is on objects made by humans for practical reasons. Without further qualification, however, this demarcation is not yet sufficiently precise to capture our subject matter, since the notion of object may be taken in a very general sense, including not only material (physical) but also abstract objects. In that case also social objects may qualify as technical artefacts since many social institutions are human made for solving practical purposes. This leads to a very broad notion of technical artefacts and of technology, including for instance bureaucracies (Pitt 2000). In this book I opt for a more narrow interpretation of technical artefacts as material (physical) means that people make and use for solving practical problems. So, the kind of technical artefacts I am interested in may be characterized as human-made physical objects for solving practical problems.[4]

Within the genetic account of technical artefacts the notion of *making* (or *creating*) plays a crucial role. However, what does it mean to make something, more in

[1] The term 'theory' stems from the Greek 'θεωρειν', of which one of its original meanings is to be a spectator at a theatre.

[2] In many practices involving scientists and engineers these two attitudes to the world may merge to such a degree that it becomes difficult to characterize such practices as either scientific or technological. Nevertheless, the distinction between these two attitudes makes sense; it is easy to point out scientific and technological practices that reflect these attitudes in more or less pure forms.

[3] For discussions of the notion of artefact, see for instance Dipert (1993), Simons (1995), Hilpinen (2004) and Margolis and Laurence (2007).

[4] This means that software programs fall outside the scope of this book. I consider software programs to be 'incomplete' technical artefacts; only in combination with the appropriate hardware that executes software programs are they able to fulfil their technical function.

particular, what does it mean to make a technical artefact? What kind of human activities are involved? With regard to these questions it is interesting to observe that artefacts in general, of which technical artefacts are a subclass, are often characterized by philosophers as creations of the human mind (Thomasson 2007, p. 52): "It is frequently observed that artifacts and other social and cultural objects are in some sense 'creations of the mind', depending in certain ways on human beliefs or activities."[5] According to this line of thinking human intentions play a crucial role in making artefacts. Physical activity alone is not sufficient to make an artefact; it may even not be necessary. Think of *art trouvé*; an object may become a piece of art solely by the intentions (mental activity) of an artist.[6]

With regard to making technical artefacts, intentions also appear to play a crucial role. In our characterization of technical artefacts as human-made physical objects for solving practical problems, these intentions are hidden in the term '*for*'. Somebody, who for no practical reason ('just for fun') sharpens one end of a wooden stick, has not made an object for doing something and thus has not made a technical artefact. The same person, however, performing the same physical actions with the intention of making a wooden stick with a sharp end for hunting and killing an animal does produce a technical artefact (a spear). Analogous to the *art trouvé* case, it may even by questioned whether physically making something is necessary for making a technical artefact. Human intentions may be considered to be sufficient for making a technical artefact out of an object found in nature. According to this line of thought, natural objects may be made into technical artefacts by human intentions alone (by using them intentionally for practical purposes; '*technologie trouvée*'). After all, what is the difference between a piece of flint found in nature and used as a hand axe, and a piece of flint brought into the same physical shape and used as a hand axe? Why call the latter a genuine technical artefact and the former not?

The view that technical artefacts are primarily or only creations of the human mind downplays or ignores the role of physical actions involved in making technical artefacts. In my opinion such a view cannot provide an adequate account of the making of modern day *engineered* technical artefacts, which will be my main focus. Most technical devices used in everyday life, such as bicycles, cars, printers, incandescent lamps, vacuum cleaners, coffee machines and so on, belong to this category. They differ from natural pieces of flint used as hand axes in that the physical objects involved are not lying around in nature to be appropriated by human beings as technical artefacts by a simple act of the mind. They have to be designed and physically made. Matter has to be transformed such that the resulting physical construction has certain capacities or shows a particular kind of behaviour. Often that is an arduous process which may involve many problems, setbacks and failures. It is not sufficient to have the brilliant idea of a phonogram; for that idea to become a technical artefact

[5] This quote is from an edited volume with the title *Creations of the mind: essays on artifacts and their representations* (Margolis and Laurence 2007).

[6] In *art trouvé* also technical artefacts may be turned into works of art by the thoughts or intentions of artists; think of Duchamp's *pissoir*.

it has to be realized, and since there are no physical objects with the capacity of reproducing sound lying around, a physical construction with that capacity has to be invented and made. Ideas are important but have to be 'materialized' or 'embodied'. The making of these engineered technical artefacts involves not only mental work, but also physical work. In technology intentions (ideas) are important but not sufficient. Intentions alone have not put humans on the moon; there are no natural objects around that may be turned into a moon rocket simply by our intentions. In the technical world intentions (ideas) have to be put to work effectively in and through matter. From a genetic point of view, therefore, engineered technical artefacts are better characterized as 'creations of mind and hand': both mental work (an inventive idea) and physical work (shaping matter/making physical constructions) are necessary in order to create them.

Instead of characterizing technical artefacts by focussing on how they come into being and on the role of physical and mental activities therein, technical artefacts may also be characterized as a special class of objects on the basis of their practical 'for-ness': a knife is for cutting, a car for driving and a copying machine for copying and so on.[7] It is this practical for-ness feature that distinguishes technical artefacts from physical objects such as a piece of flint and electrons; they have no practical for-ness or for-ness in general (a piece of flint is not for cutting, an electron is not for hitting upon a luminescent substance in a Cathode Ray Tube). There is at least one class of natural objects and phenomena that also exhibit for-ness, namely biological traits such as the organs and behavioural patterns of animals: wings are for flying and a mating dance for attracting a partner. However, this biological for-ness appears to be different from the practical for-ness of technical artefacts. The latter is intimately related to the notion of intentional use; "a knife is for cutting" roughly means "a knife is to be used for cutting". This is not the case for biological for-ness (my heart is for pumping blood, but am I to use or do I use my heart for pumping blood?).[8] As far as works of art are concerned, in so far they are also characterized by a certain for-ness, it seems questionable that this for-ness is of the same kind as the practical for-ness of the technical artefacts mentioned above. At least some social objects, like bureaucracies, may be taken to have a practical for-ness similar to technical artefacts, but as abstract objects they are different from technical artefacts. Thus, if technical artefacts are taken to be physical constructions with a practical for-ness, then that conception sets technical artefacts roughly apart from physical, biological, artistic and social objects.

From this for-ness perspective, technical artefacts may be characterized equally as physical constructions with a technical (practical) *function*. Thus, a technical artefact is always more than simply a physical structure. It is a physical structure with a for-ness which is captured by its technical function. This 'functional'

[7] As far as I know the term "for-ness" was coined by Dipert in a paper he presented at Delft University of Technology, (Dipert 2000); see also Romano (2009).

[8] Throughout this book I leave out of consideration biological technical artefacts, such as genetically modified organisms. For a discussion of functions in the natural and artificial domains, see (Krohs and Kroes 2009).

characterization of technical artefacts brings with it problems about how to interpret the notion of technical function. On the one hand, the function of a technical artefact is intimately related to its physical structure, since it is by virtue of that structure that the technical function is performed. The physical structure and technical function constrain each other, since not any physical structure can perform a given technical function or vice versa. On the other hand, the function of a technical artefact is, as will become clear in due time, intimately related to human intentions. Without human intentions no functions and without functions no technical artefacts, but only physical structures or physical objects. Just as making a technical artefact involves mental and physical activity, being a technical artefact somehow involves both human intentions and physical structures. From this perspective, therefore, technical artefacts may be taken to be creations of the mind and from matter.

So, on both the genetic and the for-ness account, the characterization of an object as a technical artefact involves an appeal to mental and physical elements. It is this combination of 'mind and matter' that makes it so difficult to fit the notion of technical artefacts into either one of the two main conceptual frameworks in use for describing parts or aspects of the world, namely the physical and intentional conceptual frameworks. The physical one, encountered in the physical sciences, conceives of the world as consisting of physical objects that interact physically. The intentional framework, underlying roughly the humanities and social sciences, conceives of parts of the world as consisting of agents, primarily human beings, who intentionally represent the world and act on it on the basis of reasons. In so far as technical artefacts are physical objects they fit into the physical conception of the world; their physical properties can be accounted for in this view. That, however, is not possible for their functional properties (their for-ness feature), since the functional properties of technical artefacts refer to intentions and intentions have no place in the physical conception of the world. In so far as they have intentionality-related functions, technical artefacts fit into the intentional conception; in this conception their functional properties can be accounted for by relating them to human ends and purposes. However, the intentional conception does not offer the resources to account for the physical features of technical artefacts.

With regard to these two conceptual frameworks technical artefacts may be said to be objects with a *dual nature* in the sense that these two different conceptualizations are necessary to account for what kind of objects technical artefacts are. The physical conceptualisation may account for the way the artefact works in terms of physical processes. But as a mere physical object, it is not a technical artefact. Without its function, the object loses its status as a technical artefact. The intentional conceptualisation may account for the function of a technical artefact in terms of what it is for, and relate this function to the realization of human ends. But a functional (means-end) description of a technical artefact effectively black-boxes its physical structure. Somehow the physical and intentional conceptual frameworks have to be combined in order to account for the specific dual nature of technical artefacts. They are hybrid objects combining physical and intentional features.

One of the main aims of this book is to present an elaboration and defence of this dual-nature view on technical artefacts. The main challenge to be faced is how the physical and the intentional conceptual frameworks are to be combined in the case of technical artefacts. With regard to the development of a coherent framework for describing technical artefacts, the notion of function plays a key role. If functions of technical artefacts are seen primarily as realized in the physical objects involved, the question remains how these functions are related to the mental states of human individuals, which form the core of the intentional conceptualisation. If, on the contrary, functions are seen primarily as patterns of mental states and exist, so to speak, in the heads of designers and users of technical artefacts, it becomes mysterious how a function relates to the physical substrate of a particular artefact. The notion of function appears to be a kind of 'bridge-concept' between the physical and intentional conceptualizations of the world since the function of a technical artefact is closely related to its physical structure on the one hand, and to human intentions with regard to that artefact on the other. That is the reason why the elaboration of a theory of technical functions plays such a prominent role in the first part of this book.

One issue, though, still has to be addressed briefly before I can go over the content of that part and of the remainder of this book. How is the notion 'practical' to be understood? I have been using that notion frequently up till now in expressions like 'practical needs and ends', 'practical purposes', 'practical for-ness' and 'practical problems'. Although it may be intuitively clear what is meant with the term 'practical', it is certainly not easy to explicate its meaning in general terms. One reason I use this notion is to connect technical artefacts and technology to problems involving practical rationality ('What to do?') and not so much to problems involving theoretical rationality ('What to believe?'). This use of the term 'practical' is in line with our characterization of the distinction between science and technology. However, this may be misleading since solving problems about what to do and what to believe may both involve the making and use of technical artefacts. Just think of the making and use of calculating equipment for scientific purposes and the complicated technology needed to settle the issue about believing whether the Higgs boson exists or not. So, a theoretically oriented endeavour like science may involve practical problems that are solved by making technical artefacts. Conversely, solving technical problems may involve solving theoretical problems. Practical problems, the solution of which may involve the making and using of technical artefacts, may turn up not only in technology and science, but with regard to almost every aspect or domain of human activity, including the arts and even morality, whether or not these activities are guided by practical, theoretical or some other form of rationality. As we remarked at the start, the making and use of technical artefacts pervades almost all aspects of human life.

I will not attempt to clarify the notion 'practical' by further opposing it to the notion 'theoretical', since that would require an in-depth examination of theoretical and practical rationality and their differences.[9] What is, I think, important with regard to the notion 'practical' in expressions like 'practical ends', 'practical

[9] For a discussion of theoretical and practical rationality in relation to engineering design, see (Kroes et al. 2009).

problems', 'practical reasons' et cetera is that this notion is somehow related to means that are primarily evaluated on the basis of their effectiveness and efficiency in realizing ends. Therefore these means fall squarely within the province of instrumental rationality. Technical and many social artefacts are such means; they are part of solutions to practical problems and are therefore evaluated on the basis of their effectiveness and efficiency. For that reason they are commonly taken to be different from works of art, that are assessed on aesthetic criteria. So, this intimate relation between the practical and instrumental rationality may distinguish technical artefacts from works of art, but not from the other close cousin, that is, social artefacts. As we have already observed, this may be achieved by taking the nature of the artefacts involved, material versus abstract, into account. However, it may be rather problematic to distinguish the technical from the social in solutions to practical problems; I will return to this topic in section 2.1.

So, this book is about objects of a particular kind that play a role in the solution of practical problems, namely technical artefacts. Its main aim is to clarify the nature of this kind of object, especially by comparing them to physical objects and social artefacts. As physical objects with a function or for-ness they appear to have a dual nature because they have features that link them closely to physical objects and to social objects. Chapter 2 sets the stage for the exploration of the dual nature of technical artefacts. On the basis of an analysis of how engineers describe technical artefacts I argue that they conceive of technical artefacts as objects with physical and functional properties. This by itself does not yet imply that technical artefacts have a dual nature. Whether that is the case, depends on the interpretation of the notion of function. As a first entrance into the domain of theories of functions I sketch two opposed positions. One, rather akin to engineering practice, couples functions primarily to physical capacities, the other, rather dominant among philosophers, couples functions primarily to human intentions. If functions of technical artefacts are equated to their physical capacities, then technical artefacts are simply physical objects and all of their properties, including their functional ones, can be captured in the language of the physical sciences. On this view technical artefacts have no dual nature. However, this view runs into great difficulties, such as dealing with the fact that technical artefacts may malfunction. If functions of technical artefacts are interpreted in terms of human intentions, then technical artefacts become objects with a dual nature in the sense that they have physical and intentional features. A problem with this line of interpreting functions is that it cannot account for how the physical and intentional features of technical artefacts are related to each other. Whether some object is, for instance, a screwdriver is, according to this line of thought, only a matter of human intentions. That is a serious shortcoming. The function of a technical artefact has to be realized by its physical structure, which implies that physical structure and technical function constrain each other. So, what is needed is an interpretation of technical functions that ties them to physical structures *and* to human intentions. Only such a 'hybrid' theory of technical functions will be able to provide a fruitful explication of the dual nature of technical artefacts.

That is the reason why I turn, in the next chapter, to an examination of some of the most important theories of technical functions discussed in the literature; my aim is to see whether there is an acceptable hybrid theory among them. In order to assess these theories it will be necessary to do some preliminary work. First, I will introduce a distinction between two kinds of functional properties that may be attributed to an object x, namely "for φ-ing" ("x is for driving screws") and "a φ-er" ("x is a screwdriver"). Generally speaking, an object x may have the functional property "for φ-ing" without having the functional property "a φ-er", that is, an object x may be for driving screws without being a screwdriver (without being an instance of a particular technical artefact kind). This is a crucial distinction that involves the distinction between accidental and proper functions. The question how these two different functional properties are related puts on the agenda the relation between theories of technical functions and theories of technical artefact kinds. I will argue that failure to take into account the distinction between these two functional properties, and the associated issue of the relationships between theories of technical functions and theories of technical artefact kinds, causes serious trouble for theories of technical functions. Second, in order to clarify what it means for functions to be mind-dependent (dependent on intentions) I discuss the general forms of epistemic and ontological theories of technical functions. This distinction is seldom made, which, I maintain, leads to confusion about the role of the mind in mind-dependent theories of technical functions. This confusion is caused by the fact that function ascriptions, as an epistemic activity, are mistaken for function assignments, as a pragmatic activity. I argue that only function assignments can ground mind-dependent theories of functions. Against the background of these distinctions I discuss the theories of technical functions developed by Searle, Preston and Houkes and Vermaas. I argue that for various reasons none of these theories is the hybrid theory of technical functions I am in search of. Each one of them fails for different reasons to explicate and to do justice to the dual nature of technical artefacts.

That leaves me in chapter 4 with the burden of developing a hybrid theory of technical functions that may account for the dual nature of technical artefacts. As a preliminary step a clarification of the normativity associated with technical functions is in order. This normativity is usually seen as a touchstone for theories of technical functions. I argue that there is nothing intrinsic about the normativity of technical functions. On the contrary, the normativity of technical functions finds its origin in the normativity associated with intentional human behaviour. Explaining the normativity of technical functions should therefore pose no challenge for a hybrid theory of technical functions, since hybrid theories interpret functions partly in terms of human intentions. Then I will argue why theories of technical functions stand in need of being complemented with theories of technical kinds and of technical artefact kinds. Technical kinds are functionally defined, whereas technical artefact kinds are defined by functional and structural (physical) features. There is a widespread (implicit) assumption that the notion of proper function, as defined in theories of technical functions, also defines membership of technical kinds. If an object has the proper function to drive screws, it is taken to be an instance of the technical kind 'screwdriver'. That assumption, however, is not compatible with the theories of technical functions discussed. This is precisely one of the reasons why these theories fail. In order to solve these problems I propose to distinguish between two kinds of proper functions, namely kind-proper functions and use-proper functions. Kind-proper functions are defined by theories of technical artefact kinds and an object that has a certain kind-proper function is, by definition, an instance of the corresponding technical kind. In order to develop a theory of technical artefact kinds, I turn to Thomasson's theory of artefact kinds, according to which an object is an instance of an artefact kind K just in case it is the result of a largely successful execution of a largely correct substantive idea of what it means to be a K-er. Elaborating this theory for technical artefact kinds, I argue that kind-proper functions are part of the 'largely correct substantive idea' of the process of creating new technical artefact kinds. Use-proper functions are defined by theories of technical functions based on use practices. Most theories of technical functions discussed in chapter 3 favour a definition of proper function on the basis of use. I argue that it is to be expected that the kind-proper function of a technical artefact will coincide typically with its use-proper function, because the technical artefact is used for what it was created for in the first place. In certain circumstances kind- and use-proper functions, however, may come apart.

The distinction between kind- and use-proper functions is one of the cornerstones of the combined theory of technical functions and technical artefact kinds that I propose. A basic assumption underlying my approach is that the making of technical artefacts, in contrast to using them, introduces new objects into the world. Barring exceptional cases of creative use, it is through the design and making of new technical artefacts (whether new instances of an already existing kind, or a first instance of a new technical artefact kind) that changes into the furniture of our universe come about. So, in order to understand the nature of technical artefacts we have to turn to theories of technical artefact kinds, in particular to theories of kind-proper functions, since the kind-proper function of an object makes it into an instance of a particular technical artefact kind. According to the theory of technical artefact kinds presented here, the kind-proper function of an object is grounded in human intentions as well as in physical features of that object. This means that this theory is a 'hybrid' theory of functions of the sort that I argued is required in order to account for the dual nature of technical artefacts.

In chapter 5 I turn to an analysis of engineering design, the starting point of the process that leads up to the actual production of technical artefacts. When we look at how the production process of technical artefacts has been institutionalized in modern industry, one of the most striking features is a strong division of labour between two forms of work, mental and physical. The mental work of conceiving of a new technical artefact kind takes place in the design phase and is done by design and development engineers, whereas the physical making is done by manufacturing engineers. This division of labour reflects the two elements that play a key role in the definition of technical artefacts (kinds) presented in the previous chapter. Design engineers work out the 'largely correct substantive idea' whereas manufacturing engineers see to it that this idea is executed in the right way.[10] As a result of this

[10] There is not always a strict division of labour; design engineers, for instance, may make real technical artefacts such as prototypes to perform experiments on.

division of labour, the outcome of the engineering design phase is a design, which is not yet a full-blown technical artefact but a blue-print of it. I address issues related to both design as a verb and as a noun. As to design as a verb, I maintain that designing may be characterized as a process in which a function is 'translated' into a structure, which means that in the design process the dual nature of technical artefacts is reflected. A brief analysis of 'means-end' reasoning is presented since this kind of reasoning appears to play a crucial role in solving engineering design problems. Our attempt at an explication of the notion of design as a noun, in expressions such as the design of this car, shows that this is a rather elusive notion. What do we mean by the design of a technical artefact, and what do we mean by saying that a technical artefact is the 'embodiment' or 'material realization' of a design? These issues lead us back to some of the topics discussed in the first three chapters. The design of a technical artefact is usually taken to be one of its defining features; an object is an instance of a particular technical artefact kind because of its design. Furthermore, describing technical artefacts as the embodiment or material realization of a design is very much in line with the dual-nature conception of technical artefacts. Nevertheless, the notion of design as a noun, although it plays a key role in engineering practice, remains a vague notion. A clear conceptual analysis of this notion is still lacking. I also analyse why the formal representation of the functional features of a design or a technical artefact turns out to be much more problematic than a formal representation of its physical features. The reason for this is to be found in the mind-dependence of functional features.

In chapter 6 I turn to the moral status of technical artefacts. This is a topic that has been much debated in recent decades among philosophers of technology and STS scholars. It is not controversial to claim that the use of technical artefacts has moral significance and that their use may be evaluated in a moral way. This moral significance finds its origin not in the technical artefacts used but in the ends pursued by humans through their use, ends that may be subjected to moral assessment. But what may be said about the moral status of technical artefacts by themselves, independent of their use? In answering this question we appear to get caught up in a dilemma. On the one hand, we may deny that technical artefacts have moral significance by themselves. This means that we end up defending some variant of the moral-neutrality thesis: technical artefacts are morally neutral instruments. In so far technical artefacts give rise to moral issues, this only occurs within their context of use. Nevertheless, it is beyond doubt that technical artefacts have a far-reaching influence on human beings; they not only influence but condition human behaviour, by making certain forms of behaviour possible or even necessary, forms of behaviour that may be morally significant. How is this deep influence to be reconciled with the idea that technical artefacts are mere morally neutral instruments? On the other hand, we may affirm that technical artefacts by themselves have moral significance and may be evaluated in a moral sense. One way to argue for this position is to point out that technical artefacts by themselves may 'harbour' or 'embody' morally significant values such as safety or privacy. Another way is to attribute some form of agency to technical artefacts that may account for the fact that technical artefacts *do* all kinds of things to human beings. This agency of technical artefacts may be taken to be morally significant in analogy to the agency of human beings. Such positions, however, incur the burden of explicating what it means for a technical artefact to 'harbour' or 'embody' by itself morally relevant values or the have some form of agency. What does it mean to claim that technical artefacts may act and that this action may be morally assessed? I argue that we may avoid this dilemma by being careful about what is meant with the notion of technical artefacts *by themselves*. If the notion 'by themselves' is construed as independent of human intentions, then the idea of technical artefacts by themselves is meaningless, since the dual nature of technical artefacts implies that human intentions are necessarily involved in what it means to be a technical artefact. However, if the notion 'by themselves' is construed as '*qua* technical artefacts', then there is room for attributing moral significance to technical artefacts by themselves, because their functional features are tied to human ends that may be morally significant. Since these functional features are also necessarily involved in what it means to be a technical artefact, we are dealing here with a form of moral significance that belongs to technical artefact themselves, *qua* technical artefacts. In other words, the world of technical artefacts, as a world of human making, 'embodies' ends that may be morally significant, but this moral significance is ultimately grounded in human agency.

In the final chapter, the Epilogue, I draw attention to socio-technical systems. These are systems that perform their overall function on the basis of the functioning of technical and social subsystems. Socio-technical systems also have a hybrid nature, but a brief examination of their hybridity shows that it is different from the kind of hybridity we encountered with regard to technical artefacts.

References

Dipert, Randall R. 1993. *Artifacts, Art Works, and Agency*. Philadelphia: Temple University Press.

Dipert, Randall R. 2000. Doing and Making: the proper place of actions and artifacts in philosophical theory. In *The dual nature of technical artefacts: kick off meeting*. Delft, Delft University of Technology (June 2000).

Hilpinen, Risto. 2004. Artifact. In *The Stanford Encyclopedia of Philosophy (Fall 2004 Edition)*, ed. Edward N. Zalta. (editor): http://plato.stanford.edu/archives/fall2004/entries/artifact/.

Kroes, Peter, Maarten Franssen, and Louis Bucciarelli. 2009. Rationality in Design. In *Philosophy of Technology and Engineering Sciences*, ed. Anthonie Meijers, 565-600. Elsevier.

Krohs, Ulrich, and Peter Kroes. 2009. Functions in biological and artificial worlds; comparative philosophical perspectives. Cambridge, Massachusetts: The MIT Press.

Margolis, Eric, and Stephen Laurence. 2007. Creations of the mind: theories of artifacts and their representations. Oxford: Oxford University Press.

Pitt, Joseph C. 2000. *Thinking about technology; foundations for the philosophy of technology*. New York: Seven Bridges Press.

Romano, Giacomo. 2009. Being for: A philosophical hypothesis about the structure of functional knowledge. In *Functions in biological and artificial worlds; comparative philosophical perspectives*, eds. Ulrich Krohs, and Peter Kroes, 167-184. Cambridge (Mass.): MIT Press.

Simons, Peter. 1995. Artefact. In *Companion to metaphysics*, eds. J. Kim, and E Sosa, 33. Oxford: Blackwell.

Thomasson, Amie L. 2007. Artifacts and human concepts. In *Creations of the mind: essays on artifacts and their representations*, eds. Stephen Laurence, and Eric Margolis, 52-73. Oxford: Oxford University Press.

Chapter 2
Technical artefacts

2.1 Demarcation

Technical artefacts come in many shapes and sizes: from rather primitive (prehistoric stone axes) to very sophisticated (modern-day laptop computer), from small (microscopic electric motors) to large (drilling platforms), from components (transistor) to whole products (calculator), from non-living (drugs) to living (genetically modified bacteria), from 'monolithic' (a screw) to systemic (infrastructure systems of various kind), from matter/energy-processing (oil refinery) to information-processing (abacus), technical artefacts with software (computers) and without (hammer), chemistry-based technical artefacts like paints, glue, fuels etc. etc. All of these have in common that they are objects, substances or systems produced by human beings for a purpose. They therefore all belong to the class of artefacts (Hilpinen 2004, p. 4–5). What is more or less specific for technical artefacts is that they serve practical purposes as opposed to artistic or other kinds of purposes and that they are typically material objects as opposed to abstract artefacts such as poems, laws, organizations or contracts.

I will limit the scope of my inquiry to paradigmatic cases of material technical artefacts such as screwdrivers, telephones, cars, staplers, lamps, printers, clocks, and so on. These paradigmatic examples all have in common that they are based on intelligent human design. By limiting myself to this type of technical artefacts, I want to avoid getting caught up in two difficult demarcation problems, one concerning the boundary between natural and technical objects, the other between technical and social objects.[1]

[1] In the following pages I will focus on the boundary between technical artefacts and natural objects. For reasons that will become clear in section 2.3, which contains a discussion of the relation between the notions of natural and physical objects, I will concentrate in the rest of this book on a comparison of technical artefacts to physical objects. Note that just as there is no sharp demarcation between technical and natural objects, there is also no sharp demarcation between technical and physical objects.

P. Kroes, *Technical Artefacts: Creations of Mind and Matter: A Philosophy of Engineering Design*, Philosophy of Engineering and Technology 6,
DOI 10.1007/978-94-007-3940-6_2, © Springer Science+Business Media B.V. 2012

Natural objects differ from technical artefacts in that they are not made by humans; they come into being without human intervention. Although it is quite common to oppose natural objects to technical artefacts, it is far from obvious that a neat demarcation line can be drawn between the two kinds of objects. What kind of human work and how much of it does it take to change a natural object into a technical artefact, to change a stone into an axe or a cow into a small milk factory? There appears to be a continuous spectrum of things running from the domain of the natural to the technical.[2] That does not mean that it follows that there is no difference between the two kinds of things. That would be tantamount to committing a fallacy of the heap. My limitation to paradigmatic examples of technical artefacts allows me to bypass largely discussions about natural-artificial borderline cases.

At first sight it seems much easier to draw a sharp boundary line between technical and social objects; examples of the latter are laws, state-borders, marriage, money, organizations, agreements, contracts etc. These formal and informal social objects (institutions) influence and govern the behaviour and cooperation of people. Like technical artefacts, many social objects are also based on an intelligent human design and serve a purpose or function. Nevertheless, they are not considered to be technical artefacts here because they are not material objects. Although social objects may involve physical objects or phenomena, for instance a contract may involve a written document or spoken words, they do not perform their social function by virtue of the physical particulars of these objects or phenomena. That is why we appeal to lawyers and not to engineers when we draw up a contract. So the most obvious way to demarcate technical from social objects appears to be by the role of material properties in the performance of social functions.

For two reasons the idea that there is a sharp boundary line between objects or systems from the domain of the technical and from the social is problematic. First, also in this case it may be argued that there is a seamless transition between the two domains. Second, some objects or systems appear to be technical and social at the same time.

As regards the first point, let me quote Searle (1995, p. 39–40) at length:

> Consider for example a primitive tribe that initially builds a wall around its territory. The wall is an instance of a function imposed in virtue of sheer physics: the wall, we will suppose, is big enough to keep intruders out and the members of the tribe in. But suppose the wall gradually evolves from being a physical barrier to being a symbolic barrier. Imagine that the wall gradually decays so that the only thing left is a line of stones. But imagine that the inhabitants and their neighbors continue to recognize the line of stones as marking the boundary of the territory in such a way that it affects their behavior. For example, the inhabitants only cross the boundary under special conditions, and outsiders can only cross into the territory if it is acceptable to the inhabitants. [...] The line of stones performs the same function as a physical barrier but it does not do so in virtue of its physical construction, but because it has been collectively assigned a new status, the status of a boundary marker.

[2] According to Preston (1998, p. 253; 2008, p. 28) there is from the point of view of using existing things for new purposes (that is, conferring a new function onto existing things, what is called in biology "exaptation") no significant difference between using natural things or technical artefacts and from that point of view "the realm of artifacts is a seamless extension of the realm of naturally occurring things."

Table 2.1 Various kinds of solutions to practical problems

Solutions to practical problems	With technical elements	Without technical elements
With social elements	Socio-technical solution	Social solution
	Traffic lights	*Traffic regulator*
Without social elements	Technical solution	–
	Fully automated car	

Here, the wall is a technical artefact, the line of stones *in so far it is the marker of a boundary* a social artefact. The physical particulars of the social object 'boundary marker' do not matter much; instead of the line of stones, it may also be a line painted on the ground. It may not even be a visible marker at all, but information stored in a geo-information system that may be consulted with some device (for instance, the GPS-system). What matters is that there is something that symbolizes or represents the boundary and may be recognized as such. In the intermediate cases between the original wall and the symbolic marker it is not clear whether we are dealing with a clear-cut technical or social object. The crumbling wall, I suppose, performs its function partly on the basis of its physical properties and partly on the basis of it being collectively recognized as a boundary marker. There is no 'natural' point at which the technical artefact turns into a social one. Only the end points can be classified unambiguously.

The second reason why a clear demarcation of the domain of the technical and the social is problematic concerns systems that perform their function on the basis of technology *together with* social rules or institutions. In order to illustrate this, consider the various kinds of systems that may be put in place to solve the practical problem of regulating traffic at a crossroads (see Table 2.1). To begin with, it may be solved in a social way, without making use of any physical technology, by introducing and enforcing rules for the behaviour of car drivers at cross roads ("give way to cars from the right"). Another social solution consists of placing a traffic regulator (traffic policeman) at the cross roads, who is invested with the authority to regulate the behaviour of car drivers. For that social solution to work it will be necessary that car drivers are able to recognize that the person at the cross roads is in fact a traffic regulator, that is, performs a social role. This may be achieved by a particular outfit that signifies that somebody wearing this outfit operates as, or counts as, a traffic regulator. These ways of solving the practical problem work by introducing social systems that consist of complex network of rules and persons in particular roles; these social systems are intended to govern and coordinate the behaviour of car drivers at cross roads. The success of these solutions depends crucially on the collective acceptance of these rules.

Another way to solve the same practical problem is not to rely on people following rules but on technical means. In the near future it may be possible to introduce technical systems, including cars equipped with fully automated vehicle guidance systems, that make it possible to regulate car traffic at cross roads in a completely automatic way, without any actions required by people in whatever role, be it the role of car driver or traffic regulator. Once the technical system has been put in place and in operation, it regulates (or better controls) the traffic flow at the cross roads all by itself. It does so with the help of sensors, activators, complex feed-back systems, information systems et cetera, whose behaviour is governed not by rules but by causal (or statistical) laws. No regulators or operators of technical devices within the system are required. The success of this kind of solution only depends on how well the technical system is functioning. No collective acceptance of rules is involved because all social roles have been eliminated. Of course, people as users of cars in order to move around have not been eliminated from the scene. They have been eliminated only in so far they operate cars, that is, perform the social role of car drivers; a fully automated car has no need for someone who operates or drives it.

Finally, our practical problem may be solved by introducing systems that combine elements from the social and technical ways. In the case of a fully automated traffic lights system, the person with the role of traffic regulator is replaced by a piece of technical hardware; the only operators that are part of this system are car drivers. There is no real technical counterpart of the issue of recognizing the authority of a traffic regulator; it makes no sense to say that a piece of technical hardware has the authority to regulate traffic. What has to be in place, though, is a system that enforces sanctions on those who do not follow the rules regarding traffic lights. In this case, a combination of rules intended to govern the behaviour of car drivers and a causally operating technical piece of hardware performs the function of regulating car traffic and both have to function well and have to be well-attuned to each other in order for the whole system to function well. Clearly this traffic lights system, taken as a solution to the practical problem of regulating car traffic, is neither a technical artefact nor a social artefact. It is a hybrid solution to a practical problem, and the system installed is a socio-technical system that combines technical and social elements.[3]

So, some practical problems may be solved by means ranging from social, through socio-technical to technical ones. In general, what kind of technical and social means are available for solving a practical problem will depend heavily on the kind of problem involved, the state of the art in technology and the available social means for regulating and coordinating the behaviour of people.

Note that there is an important difference between the example of the traffic lights system and Searle's example of the wall. Although we cannot classify the traffic lights system unambiguously as a technical or social system, we can precisely isolate the technical and social elements and describe their contribution to its functioning. So, we can take the piece of hardware out of the system and analyse it as a technical artefact. In the case of the crumbling wall, that appears to be much more difficult.

[3] According to Chittaro and Kumar (1998, p. 331) the functional representation of a traffic light "should establish a relation among the traffic light (specific system), the crossroads (the context), transitions among the three colors of the light (behaviour of the system), meaning of the colors for drivers (interpretation of behaviour), and regulation of traffic (purpose)." If we follow systems theory in assuming that the function of a system determines what is considered to be part of the system and what is part of its environment, then this implies that the traffic light is a system that also involves social/intentional elements. For more details about socio-technical systems, see Kroes and Franssen (2006), and chapter 7.

So, similar to the case of natural objects, it is not possible to demarcate the domain of the technical neatly and in an obvious way from the domain of the social.[4] That does not mean that the distinction between technical and social objects makes no sense in all circumstances. Again, there is no reason to fall into the trap of the fallacy of the heap. If, for whatever reasons, clear cut boundaries between natural, technical and social objects are necessary, the only way out is to draw the boundaries ourselves (Wittgenstein and Anscombe 2003, sections 68–69). For my purposes such clear cut boundaries are not necessary. As paradigmatic examples of social objects for comparison with technical objects I will take objects like a ten euro bill, a driver license, a transportation ticket, a ballot etc. The social world, of course, is much richer than these kinds of objects. The advantage of choosing this kind of social object for our project is that they are usually associated with readily identifiable physical objects; that puts me in a good position to compare later on the role of physical features in the realization of the function of technical artefacts and social artefacts.

Finally, there is one important class of technical artefacts that I will leave out of consideration, namely biological technical artefacts. These include genetically modified organisms used for technical purposes (e.g., genetically engineered bacteria for purification of sewage) but also organisms whose original biological functions through a long process of breeding have been adapted to human needs and have become technical functions (e.g., the Dutch dairy cow). The case of breeding shows that, just as it is not possible to draw a sharp dividing line between technical artefacts and natural and social objects, it is not possible to do so for biological technical artefacts and natural living organisms (see, for instance, (Longy 2009; Perlman 2009)).[5] According to Sperber (2007) animals and plants may have "simultaneously biological, cultural and artifactual functions." The rise of synthetic biology, based on the idea that it will be possible to synthesize with the help of 'biobricks' new organisms, may be the beginning of the advent of a whole new species of biological technical artefacts. In all these examples of biological technical artefacts biological and technical functions get combined. I see no principal reason why the analysis of technical artefacts presented here does not apply to these biological technical artefacts on the proviso that biological and technical functions are conceptually clearly kept apart.[6]

---

[4] Artefacts, and therefore also technical artefacts, are often characterized as social objects. For instance, Thomasson (2007, p. 52) writes: "It is frequently observed that artifacts and other social and cultural objects are…". So it would make no sense to try to demarcate technical artefacts from social objects, since they are social objects themselves. I find this view rather misleading, although I am going to argue that technical artefacts have a dual nature, one of which is social/intentional. Even if it is granted that technical artefacts are in some respect social objects, there is clearly a need to distinguish this kind of social objects from social objects like laws and organizations. It is this distinction that I have in mind here.

[5] For legal reasons (patent-law) it may be necessary in this case to draw a sharp boundary line between natural and technical objects or processes; see, for instance, (Koepsell 2009).

[6] For a brief discussion of technical artefacts, biological organisms and social institutions as functionally organized entities, and their intermediates or hybrids, see (Krohs and Kroes 2009).

2.2 Two conceptual frameworks

The example of the wall and the line of stones brings out an important difference between technical and social objects: the former perform their function on the basis of their physical make-up and physical processes, the latter on the basis of social/intentional processes. The wall keeps intruders out by its physical properties, the line of stones by being recognized by intruders as being a symbol of a boundary. It is this difference that is at first sight of great importance for understanding the distinction between technical and social artefacts. So let us pause for a moment on the distinction between physical and social/intentional processes.

Different conceptual frameworks are in use for describing physical and social/intentional entities and processes. For our purposes the following, very schematic, characterization of these two frameworks will do. A physical system is described in terms of properties like mass, geometrical form, charge, velocity, energy etc.[7] All its physical properties together determine the state of the system and changes in this state are governed by the (deterministic or probabilistic) laws of nature. The initial state of a physical system (together with possible influences on the system from the environment) determines (deterministically or probabilistically) the state of the system at later times. The initial states are the causes of later states and together with the laws of nature explain those states. The social/intentional conceptualization is used in describing and explaining human behaviour individually or collectively. It is based on the idea that human beings act on the basis of beliefs, desires and intentions. These mental states are taken to be reasons for acting and reasons play a dominant role in predicting and explaining social/intentional phenomena. But reasons and causes have different features. Reasons, in contrast to causes, may be evaluated or criticised, they justify actions (this is true even if reasons are taken to be causes of action (Davidson 1963)), and they have to be intelligible for the acting person. Moreover, human agents pursue ends and make choices and may deliberate how to act in the light of their ends. That is the reason why teleological explanations play such an important part in explaining human behaviour. It would be a category mistake, however, to apply such notions to physical objects or systems; electrons or planets don't deliberate about how to behave (about what to do). Within a social/intentional perspective rule following plays an important role in describing the behaviour of human beings. All these features make the physical conceptualisation very different from the social/intentional conceptualisation.[8]

---

[7] Which physical quantities are considered to determine the state of a system depends upon the level of description (e.g. macro- or microscopic) and the chosen perspective (mechanical, thermodynamic, electromagnetic etc.).

[8] The following quote from Dretske (2006, p. 107) illustrates this difference in terms of what he calls 'minimal rationality': "When you make a sudden movement towards my eyes, I blink. I cannot help myself. Of course, I do not want your finger in my eye. I also believe one way to keep your finger out of my eye is to close my eye when you poke at it. But though I think these things, and though I close my eye when you poke at it, I do not close my eye because I think these things. I would close my eye whether or not I had these beliefs and desires. The mechanisms for these

The existence of different conceptual frameworks for describing and explaining various phenomena in the world is by itself not a problem (except perhaps for people with strong reductionistic inclinations). Problems may arise, however, when the use of these different conceptual frameworks leads to rival descriptions for one and the same phenomenon. Probably the most well-known example of such a clash between conceptual frameworks is the mind-body problem. The raising of a hand in order to vote in a meeting may be explained in two different rival ways, namely in terms of physiological causes and in terms of reasons (mental processes). The question how to deal with these two seemingly incompatible explanations of the same phenomenon has given rise to a whole new branch of philosophy, namely the philosophy of mind. As I will argue in more detail below, also in the case of the description of technical artefacts the two conceptual frameworks meet. However, the situation is not similar to the one in the mind-body problem. In this case the two modes of description are not in competition, but complementary to each other. This means that there is no risk of mingling incompatible conceptual frameworks when I claim, for instance, that both conceptual frameworks, the physical and the social/intentional one, are necessary for describing technical artefacts. With regard to technical artefacts it is not incompatibility that we will have to worry about, but rather the danger of incoherency, that is, the danger of accepting two parallel, but *unrelated* descriptions of technical artefacts. That leads to a fragmented, 'incoherent' conception of technical artefacts, within which it will be difficult to do justice to the conception of technical artefacts as encountered in engineering practice. According to the latter, features of technical artefacts that are described within the physical conceptual framework and features that are described within the intentional conceptual framework may strongly constrain each other. If that is the case, we need to know how the two conceptual frameworks hang together.[9]

The broader context within which I will place the question about the nature of technical artefacts is how the concept of technical artefact fits into these two conceptualisations. Within this broader context our starting question – What kind of objects are technical artefacts? - may now be rephrased in the following way: *What conceptual framework is necessary for a description of technical artefacts and how does this conceptual framework relate to the physical and social/intentional conceptualization of (parts of) the world?* My strategy for clarifying the nature of technical artefacts will mainly consist of a comparison of descriptions of technical

reflexes are hard-wired. They swing into action well before thought has time to act. I have reasons to close my eyes, but my reasons for closing them are not the reason I close them. So despite the fact that I do exactly what I think will get me what I want, my behavior is not a purposeful act. It does not exhibit what I will call *minimal rationality*. Though the behavior is in conformity with thought, it is not explained, not governed, by thought. Minimal rationality requires that thought be involved in the process by means of which the behavior is produced." In fact there is in such situations no reason why I close my eyes, but only a physical cause.

[9] Franssen (2008) has argued that the social/intentional conceptualisation presupposes the physical one and that the social/intentional language is an extension of the physical language. I will leave it an open matter whether this is indeed the case (see also Searle (1995, p. 55-56)). Even if it is, the problem of the coherence of the two descriptions for technical artefacts presents itself.

artefacts with descriptions of physical and social/intentional entities in order to lay bare their similarities and differences. However, before we can proceed with this task, some issues about terminology have to be put out of the way.

2.3 Terminology

The first terminological issue concerns the use of the notions 'natural' and 'physical'. In the following I will contrast technical artefacts primarily with physical and not so much with natural objects as I have done so far. My first reason for doing so is to avoid the distinction between the natural and the artificial. It is according to Dipert (1993, Ch. 12) a rather fateful and obscure contrast. It is obscure in that it is controversial whether there is a genuine distinction to be found or to be made and if so where and on what grounds or criteria the demarcation line is to be drawn. Some deny that there is a viable distinction and argue that everything is natural ("human beings are an integral part of nature and, therefore, whatever they produce, including technical artefacts, is natural"). Others come up with demarcation criteria that may easily lead to the conclusion that everything on earth is artificial ("nature as that which is not affected by human beings"). It is a fateful distinction in that it "calls forth deep and dark interests. It strikes in us chords of how we conceive of our place and the place of creatures like ourselves" (Dipert 1993, p. 224) in the world. Since many of these issues surrounding the contrast between the natural and artificial are not relevant for my purposes, I will avoid from now on reference to the notion of the natural as much as possible.

My second reason is that the physical processes on the basis of which technical artefacts perform their function may be considered to be natural processes and that therefore the physical systems involved may be taken to be natural systems, in spite of the fact that these systems are designed and made by human beings for practical purposes. This claim needs some clarification since I started off by more or less defining technical artefacts, in contrast to natural objects, as human made physical systems with a practical purpose. So, how can these physical systems be natural systems? To see how this is possible, two different notions of natural have to be distinguished, one relating this notion to the history of an object, the other to the kind of behaviour exhibited by the object. More in detail, a human-made physical object that realizes a technical function may be conceived, analysed and described as simply a physical system and nothing more. As I will argue in more detail later on, an analysis of an object as a technical artefact has to take into account, apart from its physical features, its design and function. In a physical analysis of that object, however, there is no room for its design and technical function. It is not necessary to approach technical artefacts from what Dennett (1987) calls the design stance. We may also take the physical stance or start from the physical conceptualization and abstract from the fact that they are designed objects for achieving practical ends. If we approach a technical artefact from a physical stance, what is left of a technical artefact is simply some physical object or system of which the physical behaviour may be studied as of any other physical system. In the case of technical artefacts, this physical system is human-made, but that does not make its physical behaviour unnatural or artificial. If that were the case, modern experimental science would for the most part study artificial phenomena.[10] The fact that a physical system is the result of intentional human action, and therefore artificial on the genetic account, does not matter much from a physical stance. From that stance, any physical system behaves naturally, that is, according to the laws of nature; its history (whether it is a 'naturally' occurring system or came about as the result of intentional human action) is irrelevant since that has no influence whatsoever on its physical behaviour. From this perspective, my comparison of the description of an object as a technical artefact with the description of the 'same' object as a physical system may be taken as a comparison of a technical artefact with a natural object, in the sense of an object that behaves naturally.

A second terminological issue concerns the use of the notions 'intentional' and 'social'. In the description of the social/intentional conception of the world, I have used these two notions so far indiscriminately as if they would have the same meaning. That, of course, is not the case. My reason for treating them as interchangeable is that I do not want to enter here into a discussion about such complex issues as whether social phenomena are necessarily intentional, whether they presuppose some form of collective intentionality, which cannot be reduced to individual intentionality, whether individual intentionality is strongly conditioned by social context etc. I will keep on using the notions 'intentional' and 'social' indiscriminately as referring to situations in which individual intentions and/or intentions involving social groups are involved. Whenever it may be necessary to distinguish between individual intentions and intentional phenomena involving social groups I will do so explicitly.[11]

With these terminological issues out of the way, I can now point out two features of technical artefacts we have already encountered that *prima facie* set them apart from physical and social objects. First, what distinguishes technical artefacts from mere physical objects is that the latter have no function. Electrons, molecules or planets, qua physical objects, have no functions. Leaving out of consideration the biological sciences, the notion of function is not part of physical discourse (to be more precise, it plays no role in the description of physical reality; it plays nevertheless an indispensable role in experimental physics). But technical artefacts have a function or purpose, a feature that they share with most, if not all social institutions, such as marriage, corporations, money etc.[12] Indeed, social institutions perform

[10] Hacking (1983) has put forward the claim that physical phenomena are created and not discovered. For a critical analysis of this claim, see Kroes (2003).

[11] In my opinion technical artefacts are necessarily intentional objects, but that does not imply that they are also necessarily social objects. I leave open the possibility of Robinson Crusoe creating a first instance of a new technical artefact during his solitary stay on his island. This is in line with the theory of technical artefact kinds to be presented in chapter 4.

[12] In the following I will stick to the notion of function. For a discussion about whether the notions of function and purpose can be used interchangeably, see (McLaughlin 2001, p. 52) and (Vermaas and Dorst 2007).

functions too. Second, the way social objects perform their functions appears to be fundamentally different from the way technical artefacts do, as the example of the traffic light illustrates. Social institutions perform their function on the basis of collective acceptance of rules by humans. Technical artefacts, in contrast, perform their function on the basis of their physical make-up on the assumption that they are not broken and that they are used properly. This assumption about proper use appears to make the functioning of technical artefacts dependent on intentional states of users after all. That, however, is not the case. Whether or not the telephone on my desk works, that is, performs its function, *given that I use it properly*, does not depend in any way on my beliefs, desires or intentions with regard to this telephone. When I 'correctly use' a traffic light to cross a road, the function of the traffic light of coordinating the traffic at the cross roads will only be performed adequately if all traffic participants accept the rules involved. In this case, the performance of the function depends on the intentional behaviour of the users.

By now the stage has been set for a more substantive analysis of the nature of technical artefacts. I will follow a kind of naturalistic approach, that is to say, I will turn to engineering practice to see how technical artefacts are described by engineers and what kind of conception of technical artefacts they employ. The reason for turning to engineering practice is that engineers are experts in designing, making, analysing and describing technical artefacts and so their way of describing and conceiving technical artefacts may be taken to be a fruitful anchor point in our quest for the nature of technical artefacts. In a similar way we turn to the descriptions of physical phenomena and the underlying conceptions of the physical world provided by the physical sciences when we are interested in the nature of the physical world. More generally, however, I appeal to everyday intuitions about how we describe and classify technical artefacts. For instance, I make use of the claim that a malfunctioning TV-set with a broken on/off switch is still a TV-set or of the claim that an object, such as a coin, used as a screwdriver is not a screwdriver. These claims, I think, are in accordance with how we think about technical artefacts in everyday practice. For the moment I interpret such claims not as statements about the ontology of technical artefacts but as statements about how in everyday practice we conceive of, describe and classify technical artefacts. So, my analysis of the nature of technical artefacts is grounded in the way technical artefacts are conceptualized in these every day and specialized practices. At the end of this chapter I will relate the outcome of this conceptual analysis to ontological issues about technical artefacts.

2.4 The structure-function conception of technical artefacts

I will examine a few examples of descriptions of specific technical artefacts by engineers, assuming that these artefacts are representative for the class of technical artefacts I am interested in. My first example is the description of the electric incandescent lamp by Edison taken from his patent application of 1879 (see Fig. 2.1).

To all whom it may concern:
Be it known that I, THOMAS ALVA EDISON, of Menlo Park, in the State of New Jersey, United States of America, have invented an Improvement in Electric Lamps, and in the method of manufacturing the same, (Case No. 186,) of which the following is a specification.
The object of this invention is to produce electric lamps giving light by incandescence, which lamps shall have high resistance, so as to allow of the practical subdivision of the electric light.
The invention consists in a light-giving body of carbon wire or sheets coiled or arranged in such a manner as to offer great resistance to the passage of the electric current, and at the same time present but a slight surface from which radiation can take place.
The invention further consists in placing such burner of great resistance in a nearly-perfect vacuum, to prevent oxidation and injury to the conductor by the atmosphere. The current is conducted into the vacuum-bulb through platina wires sealed into the glass.
The invention further consists in the method of manufacturing carbon conductors of high resistance, so as to be suitable for giving light by incandescence, and in the manner of securing perfect contact between the metallic conductors of leading-wires and the carbon conductor.....

In the drawings, Figure 1 shows the lamp sectionally. *a* is the carbon spiral or thread. *c c'* are the thickened ends of the spiral, formed of the plastic compound of lamp-black and tar. *d d'* are the platina wires. *h h* are the clamps, which serve to connect the platina wires, cemented in the carbon, with the leading-wires *x x*, sealed in the glass vacuum-bulb. *e e* are copper wires, connected just outside the bulb to the wires *x x*. *m* is the tube (shown by dotted lines) leading to the vacuum-pump, which, after exhaustion, is hermetically sealed and the surplus removed.
Fig. 2 represents the plastic material before being wound into a spiral.
Fig. 3 shows the spiral after carbonization, ready to have a bulb blown over it.
I claim as my invention –
1. An electric lamp for giving light by incandescence, consisting of a filament of carbon of high resistance, made as described, and secured to metallic wires, as set forth.
2. The combination of carbon filaments with a receiver made entirely of glass and conductors passing through the glass, and from which receiver the air is exhausted, for the purposes set forth.
3. A carbon filament of strip coiled and connected to electric conductors so that only a portion of the surface of such carbon conductors shall be exposed for radiating light, as set forth.
4. The method herein described of securing the platina contact-wires to the carbon filament and carbonizing of the whole in a closed chamber, substantially as set forth.
Signed by me this 1st day of November, A.D. 1879.

THOMAS A. EDISON.

Witnesses:
S.L. GRIFFIN,
JOHN F. RANDOLPH

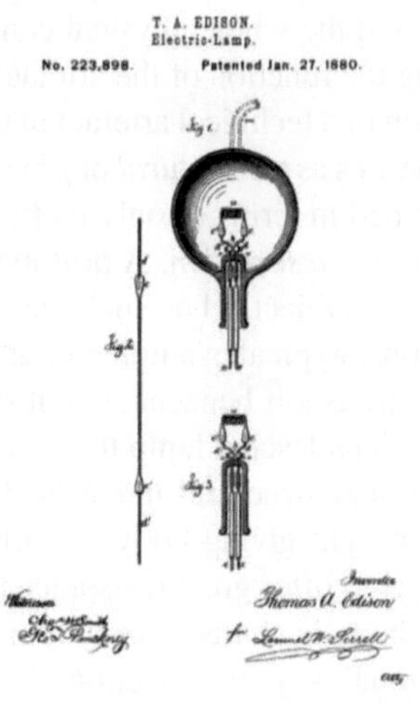

Fig. 2.1 Part of Edison's 1879 patent application for an incandescent lamp

The application starts with a description of the aim of his invention. This aim is the production of electric lamps with a high electric resistance so as to make parallel connection possible. The most innovative feature of his invention is the light-giving element made of carbon. It plays a double role: on the one hand its role is to produce light, on the other hand to restrict the electric current through the lamp.

Another important feature of this invention is that this element is placed in vacuum in order to avoid its oxidation or other forms of degradation by air. Thus, also the vacuum in the glass bulb plays a specific role.

So, the opening part of the text of the patent application presents a description of the lamp in terms of its aim or purpose. This aim refers directly or indirectly to a context of human action (namely the production of light under certain circumstances). Given this aim, Edison underlines those properties and features of the lamp that contribute to the realisation of this aim. In other words, he describes what in modern engineering terminology would be called the function of the lamp itself and of its various parts. True, he does not use the notion of function itself; in describing the lamp and its part he makes use of expressions like 'arranged in such a manner as to offer…', 'to prevent…', and 'so as to be suitable for…'. These expressions refer to aims and the parts described are the means invented by Edison to realise these aims. In modern engineering practice this boils down to attributing functions to these parts.

The description of the lamp in the second part of the text has a different character. Referring to the drawings in the application, Edison now describes the geometrical form of the lamp and of its parts and the materials from which they are made.[13] He does not mention aims or functions (except in the description of the clamps '*h h*', where he uses the expression 'serve to…'), but describes as accurately as possible 'what the lamp looks like'. Now the focus is on the material/physical properties of the lamp and its parts and the way in which these parts are physically connected to each other. In principle, anybody with the appropriate skills may be expected to be able to produce these parts and the whole physical construction on the basis of this description without knowing the function of the artefact or of its parts.

I will refer to a description of a technical artefact in terms of only its geometrical, physical and chemical properties as a *structural* or *physical description*.[14] Whenever a technical artefact is described in terms of only its function (its functional properties) I will speak of a *functional description*. A prototypical example of a structural description of an object x is: "Object x has such and such mass, shape, chemical constitution, and so on."; a prototypical example of a functional description is: "The function of x is to y (e.g., to transport benzene, to cut steel plates, to dry hair)."[15] In Edison's description of the incandescent lamp there is by no means a clear separation between the functional and structural mode of description. When he writes: "The invention consists in a light-giving body of carbon wire or sheets coiled or arranged in such a manner as to offer great resistance to the passage of the electric current…" the light-giving body is characterized both in a structural (x is made of carbon; coiled) and a functional way (the function of x is to offer great resistance).

[13] In another part of the text, not reproduced here, he describes how certain components of the lamp, such as the carbon filament, can be made.

[14] A structural or physical description of a technical artefact describes what engineers refer to as the form of a technical artefact: "The term "form" is used to relate any aspect of physical shape, geometry, construction, material, or size." (Ullman 1992, p. 20).

[15] Note that in a functional description also structural concepts may occur in the phrase replacing y; this does not imply, however, that the object x itself is described (partly) in a structural way.

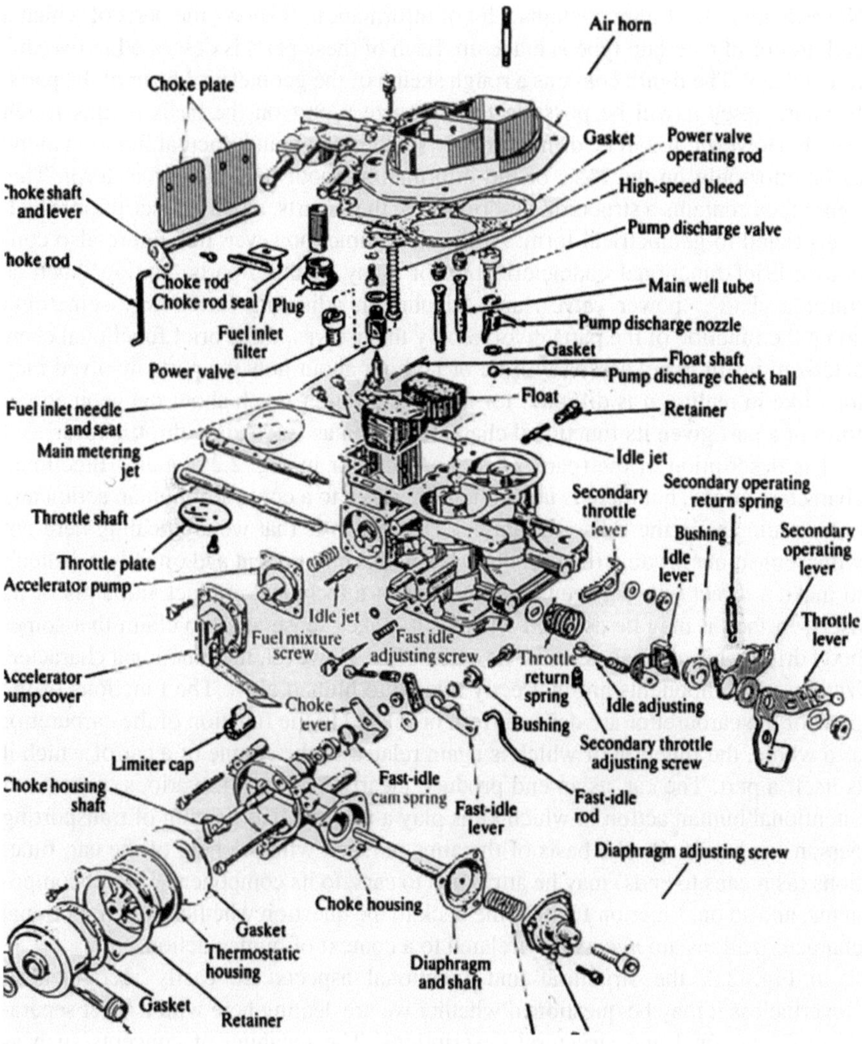

Fig. 2.2 An exploded view of a carburettor, taken from Ullman D.G. (1992), *The mechanical design process*, McGraw-Hill; reprinted with permission of The McGraw-Hill Companies

In his description, functional and structural characterizations of (parts of) the lamp are run together.

In our next example, the functional and structural elements in the description of a technical artefact are easier to distinguish and separate. Fig. 2.2 contains a so-called 'exploded view' of a car carburettor. Note, in the first place, that this figure does not present a complete description of the artefact involved (neither does Edison's description of his lamp). For instance, no information is given about the materials from with the various parts are made, nor about their size.

Nevertheless, this figure contains a lot of information. It shows the parts of which a carburettor of a certain type is made up. Each of these parts is described in two different ways. The figure contains a rough sketch of the geometrical form of the parts. In many cases it will be possible to recognize a part on the basis of this rough sketch. However, it will be difficult to say something about its actual function in the carburettor only on the basis of the information about its geometrical form. This figure thus contains a structural description of these parts, albeit a description which is restricted to geometrical form. At the same time, however, this figure also contains a brief functional characterization of many of these parts. Notions such as 'throttle shaft', 'power valve', and 'diaphragm adjusting screw' say something about the function of the parts involved. By themselves, these brief functional characterizations in many cases say little or nothing about how the parts involved may look like in reality; it is difficult, for instance, to infer much about the geometrical form of a part given its functional characterization as 'secondary throttle lever'.

The description of the (parts of the) carburettor in Fig. 2.2 contains functional characterisations, but there is no explicit reference to a context of human action and human aims as in the Edison example. The reason is that we are dealing here not with a consumer product (end product) but with a component and often it is difficult to make a direct link between a component of a technical artefact and aims of its users. In fact, it may be doubted whether it makes sense at all to claim that somebody driving his car makes use of its carburettor. However, the functional characterizations of components are indirectly related to human aims. The functions of the parts of the carburettor are derived from or related to the function of the carburettor as a whole, the function of which is again related to the engine of a car of which it is itself a part. The car, as an end product, clearly is related to various contexts of intentional human action in which aims play a role, such as the aim of transporting persons or goods. On the basis of the aims pursued with the help of the car, functions (as means to ends) may be attributed to cars, to its components, its subcomponents, and so on. Later on I will come back to the question whether these functional characterizations are *necessarily* related to a context of human action.

In Fig. 2.2, the structural and functional aspects are easily recognizable. Nevertheless it may be questioned whether we are dealing here with a strict separation of functional and structural descriptions. The meaning of concepts such as 'screw', 'shaft' and 'lever' seems to involve, apart from reference to functional properties, also reference to structural properties; for instance, a screw has a screw thread with a certain pitch, et cetera. The following two examples illustrate that engineers and technicians in certain circumstances strive for purely functional and structural descriptions of artefacts.

Fig. 2.3, adapted from Otto and Wood (2001, p. 163), presents a purely functional representation of a certain kind of technical artefact, namely a nail clipper. The most striking feature of this description is that the nail clipper itself is represented as a black box. Nothing is said about its physical structure. What is being specified is what the thing is intended to do. This is done by accurately describing the input and output of the black box (the nail clipper); what is hidden in the black box is supposed to transform the given input in the desired output. The input and

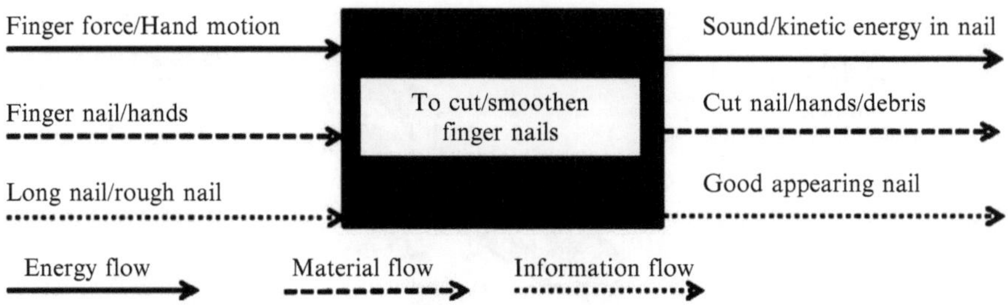

Fig. 2.3 A purely functional representation of a nail clipper

output are specified in terms of different types of flows, namely flows of matter, energy and information.

Thus, a purely functional representation of a technical artefact describes what it is supposed to or ought to do, not how this is to be accomplished (Ullman 1992, p. 20):

> A common way of classifying mechanical devices is by their function. In fact, some devices, having only one main function, are named for that function. For example, a screwdriver has the function of enabling a person to insert or remove a screw. The terms "drive," "insert," and "remove" are all verbs that tell what the screwdriver does [...] In telling what the screwdriver does, we have given no indication of how the screwdriver accomplishes its function. To answer how we must have some information of the form of the device.

In engineering practice functional representations of technical artefacts play an important role in the early phases of design processes. As I will argue in more detail later on (chapter 5), design processes may be characterized as processes in which functions are translated into structures. Especially in the design of complex artefacts, the technique of functional decomposition is used in order to structure the design task and the design process: the function of the artefact as a whole is decomposed in sub-functions, which in turn are decomposed in sub-sub-functions, and so on.

Our final example concerns a purely structural description of a technical artefact. This kind of description is the mirror image of the functional kind. Suppose that a description of a technical artefact is given that refers only to structural properties of the artefact. Nothing is said explicitly or implicitly about what it is for, i.e., about its functional properties. On the basis of knowledge of these structural properties and given sufficient knowledge about the physical laws governing the behaviour of the artefact, it is in principle possible to calculate all kinds of input-output relations for this artefact (treated as a physical system) (see Fig. 2.4). But then still the question remains which one of all these input-output relations is the desired or intended one. This question cannot be answered by simply adding more structural knowledge about the technical artefact. The reason is that in a structural description the context of human action (use), to which the function of the thing is related, is treated as a black box. In contrast to a purely functional description, which specifies what a technical artefact ought to do without describing how it actually accomplishes this, a purely structural description specifies the actual physical make-up of the artefact

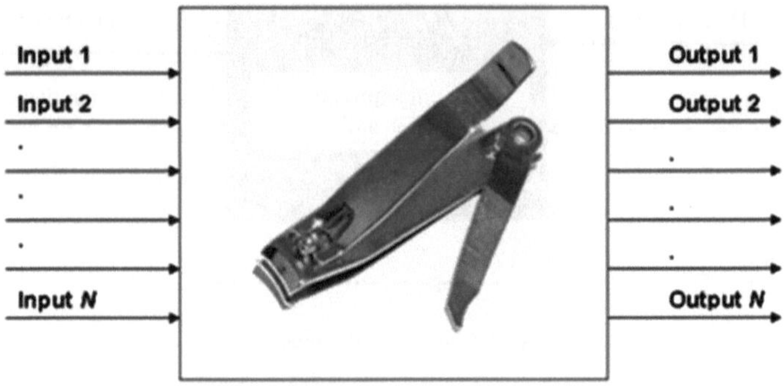

Fig. 2.4 Part of a purely structural representation of a nail clipper

and, with sufficient physical knowledge, what it is able to do, without describing what it is expected or ought to do.[16]

Purely structural representations of technical artefacts also play an important role in engineering practice. The reason is that only on the basis of a structural description of an artefact it will be possible to actually make it. As long as the description contains parts, which are characterized in a functional way, it will not be possible to produce the artefact. This means that if the design of a technical artefact is seen as a blueprint for its fabrication, then a complete structural description of the artefact has to be contained in the design. Our use of the notion 'complete' here stands in need of further clarification. It has a double meaning. In the first place it means that no functional elements are present in the description, i.e., the description has to be a purely structural one. It is, however, not sufficient to require a purely structural description. The description has to be complete in yet another sense, namely that it specifies all physical properties that are relevant for adequately performing its function. This is the second sense in which we use the term here. Note that the structural description does not have to be complete in the sense that it describes all physical properties of the object up to its last detail (many of which are irrelevant for the function of the artefact). As an example take the description of a class of standardised technical artefacts, namely hexagon nuts with flange (fine pitch thread) as contained in ISO-norm 10663. These hexagon nuts are technical artefacts of the type 'fasteners'. This is a functional characterization of this kind of object. But the description of these objects in the norm itself is of a structural kind. Part of this description is given in Fig. 2.5. The meaning of the symbols used in Fig. 2.5 and the measurement methods to be used for checking whether an object of this kind meets the required specifications are themselves specified in other norms. The values of the various symbols are specified in separate tables. All of this is to ensure that hexagon nuts satisfying this ISO-norm may be used interchangeably in engineering practice. This is achieved by

[16] In archaeology we also come across purely structural descriptions of objects that are known (or supposed) to be technical artefacts but of which the specific practical functions are unknown.

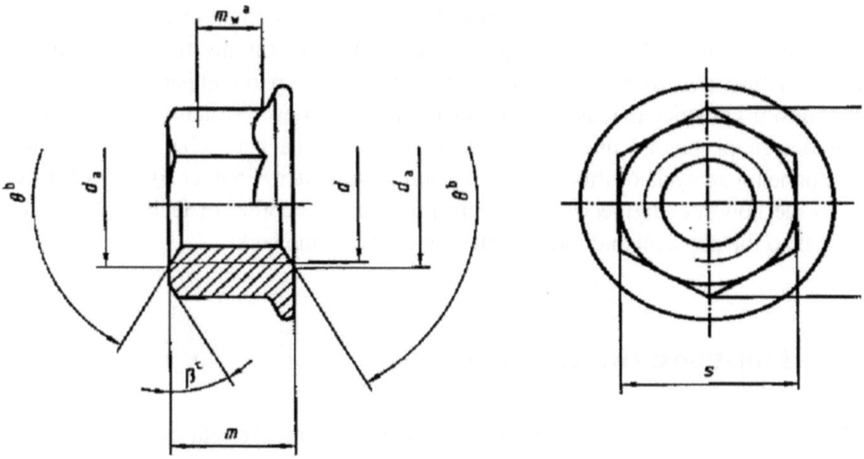

Fig. 2.5 Part of a purely structural description of a technical artefact

providing an unambiguous structural description of this kind of technical artefact, with all relevant tolerances and so on.

From these examples I conclude that engineers make use of functional and structural properties in their descriptions of technical artefacts and that the use of both kinds of properties is indispensable for engineering practice. Many descriptions have a hybrid character in the sense that structural and functional concepts are used side by side. But in certain circumstances purely functional or structural descriptions are desirable or necessary. These present only partial descriptions of technical artefacts: purely functional ones ignore physical/material aspects, whereas the structural ones ignore functional aspects. Structural and functional descriptions are not rival descriptions but complementary to each other. From an engineering perspective a comprehensive description of a technical artefact has to contain a description of all of its functional features and of all of its structural features that are relevant for performing its function. I will refer to the underlying conception of technical artefacts as the *structure-function conception*; it takes technical artefacts to be physical structures (systems) with technical functions.

The structure-function conception of technical artefacts reflects the use of two forms of knowledge in engineering practice, namely knowledge about their structural and about their functional features. From a logical point of view, these appear to be two distinct forms of knowledge. It is not clear how a structural description of a technical artefact can be derived logically from a purely functional (black box) description, in particular if it is taken into account that functions are multiple realizable. Conversely, it is logically problematic to derive a functional description from a structural one; the functional description implies claims about what a technical artefact ought to do or is supposed to do and such claims cannot be derived from a structural description without violating the alleged is-ought dichotomy (for more details, see (Kroes 2006)). From an engineering practice point of view, however, this does not mean that engineers treat these two forms of knowledge as completely independent. The structure of a technical artefact and its function are considered to be intimately related, since not any structure can perform any function. Function and structure *constrain* each other. In line with this it is part and parcel of engineering (design) practice to reason from functional properties to structural ones (in solving design problems) and vice versa (in justifying a particular design). I will postpone an analysis of this kind of reasoning until later (Chapter 5). Our next step is to see how the engineering structure-function conception of technical artefacts relates to the physical and intentional conceptual frameworks.

2.5 Functions: two story lines

The engineering structure-function conception of technical artefacts is based on two modes of representation or description, the structural and functional one. How are these two modes of description related to the physical and intentional conceptual frameworks and descriptions introduced earlier? The answer to this question determines largely how technical artefacts are related to physical and social objects, and hinges crucially on the interpretation of the notion of function. In this section I present a rough sketch of two lines of reasoning to be found in the literature, one in which functions are interpreted in terms of physical capacities and/or physical behaviour and which leads to the conclusion that technical artefacts are very much like or identical to physical objects. In the other, functions are interpreted mainly in terms of human intentions; according to this position, the function of a technical artefact is not determined by its physical structure but by human intentions and therefore a description of a technical artefact will have to make an appeal to concepts from the intentional conceptual framework. As we will see, both views on technical functions show serious shortcomings.

On the relation between the structural and physical descriptions we can be brief. A structural description of a technical artefact is essentially the same as its physical description.[17] It represents the artefact as conceived within the framework of the physical sciences. It abstracts from all those aspects of a technical artefact that do not fit into the conceptual framework of these sciences. Structural/physical properties are often characterised as intrinsic properties, i.e., as observer- or agent-independent properties. Intrinsic properties refer to features of the world that exist independently of the mental states of observers or agents; they are mind-independent. From a structural perspective, therefore, an object is described independently of what it might be used for by agents.

The relation between the functional description and the physical and intentional description is much more problematic. I start with outlining a view that tries to ground functions of technical artefacts in their physical properties, that is, a view that tries to make the functional description part of the physical description.

[17] That is the reason why I will use the terms 'structural' and 'physical' interchangeably.

2.5.1 Functions and physical capacities

This view lays much emphasis on the intimate relationship between the structure of a technical artefact and its function, since what matters is that the structure of a technical artefact has to be such that it can realize its function. Because of this, the design and development of technical artefacts is often such a difficult task. It takes much energy and intelligence to make physical objects that can perform a desired function. Years of training and experience of engineers and technicians are necessary to design, develop and produce state of the art machine tools, audio and video equipment, means for transportation, communication equipment etc. A lot of 'intelligence' and ingenuity goes into or is embodied in these technical artefacts. Their design, development and production may take many failures and struggles before an acceptable result is obtained. It is one thing to have a brilliant idea for a new kind of technical artefact based on a revolutionary design, it is another to embody that idea in matter and make it work in practice. It is in this transition that most of the 'hard' engineering work takes place. Here engineers try to mould nature according to human desires. To do so the forces of nature and the properties of materials have to be analysed, mastered and exploited in the hope of producing a physical construction that has the specific physical capacity that is necessary for performing the desired technical function. If eventually successful, the resulting physical construction is a genuine technical artefact with the specific physical capacity as its function; moreover, it is a special technical artefact, since it is the first instance of its kind. If not successful, the physical construction is just some human-made physical object that may end up in a dustbin or a junk yard. From such observations it is but a small step to assume that functions are all about physical capacities.

Also from a user perspective it may be argued that the function of a technical artefact somehow resides in its physical capacities. Among the many things on my desk there are a telephone and a stapler. What makes the one thing a telephone and not a stapler is, intuitively, its physical make-up: it has the right physical structure or capacities for making telephone calls with. This physical capacity is the function of that thing and because of this function this thing is what it is, namely a telephone.

The interpretation of technical functions in terms of physical capacities is very germane to the engineering way of thinking. Among engineers and design methodologists it is quite common to describe functions in terms of physical capacities, physical behaviour or in terms of input-output relations of systems. These input-output relations are preferably represented as mathematical functions. These mathematical functions describe the behaviour of the system that corresponds to its technical function. For the interpretation of the notion of technical function, however, it makes a lot of difference whether *actual* or *intended* (desired, required) behaviour is meant. If functions correspond to actual behaviour, then they may be interpreted as physical properties; if they are related to intended behaviour, functions become intimately tied to human intentions.[18] There is no agreement among engineers and design methodologists about whether functions correspond to actual or intended physical input-output relations. For instance, Pahl and Beitz (1996, p. 31) apply "the term *function* to the general input/output relationship of a system whose purpose is to perform a task." According to Roozenburg and Eekels (1995, p. 96) "the function of a system is the intended transformation of inputs into outputs." Chandrasekaran and Josephson (2000) distinguish between environment-centric and device-centric views on function; the former ties functions primarily to human goals and intended behaviour of the device, the latter to the physical capacities of the device. Sometimes a distinction is made between two different kinds of function, one referring to actual behaviour, the other to intended behaviour.[19] For the moment I will focus on interpretations that relate functions to actual behaviour (if functions are interpreted in terms of intended behaviour, then a different picture emerges, to be discussed in the second story line below).

Within philosophy there is also an influential view on functions that, when applied to technical artefacts, ties technical functions strongly to actual physical behaviour and capacities. That is the causal-role interpretation of functions proposed by Cummins (1975). He interprets the function of a component of a system as that physical capacity that contributes to the realization of a capacity of the system as a whole (Cummins 1975, p. 765):

> To ascribe a function to something is to ascribe a capacity to it which is singled out by its role in an analysis of some capacity of a containing system. When a capacity of a containing system is appropriately explained by analysing it into a number of other capacities whose programmed exercise yields a manifestation of the analysed capacity, the analysing capacities emerge as functions

This approach can be applied to capacities of all kinds of systems, such as capacities of biological systems (e.g., the capacity to fly) or capacities of the human mind (e.g., the ability to solve a mathematical problem). But it may also be applied to attribute technical functions to components of a technical artefact by an analysis of the capacity of the whole artefact into the capacities of its components. For instance, the capacity of the hands of a mechanical clock to perform a uniform motion can be explained in terms of the capacities of the gears, spring and other components. In the case of technical artefacts these capacities are physical capacities, or physical dispositions. The function of the component is the causal role it plays in realizing a

[18] Note that there is an important difference between the notions of a mathematical and a technical function, when the latter refers to intended behaviour. A technical function in the sense of intended behaviour has a normative element that is not present in mathematical functions: the technical artefact is supposed to (ought to) transform the input into the (desired) output. Whenever it does not do so, the technical artefact is not functioning as intended. But even when it is not performing its intended function, the input and output of a technical artefact may be represented by a mathematical function.

[19] See Rosenman and Gero (1994) and Chandrasekaran (2005) for various definitions of functions within an engineering context.

capacity of the technical system of which it is part.[20] According to this view, human intentions play no role in attributing functions to components of technical artefacts.[21]

In case the interpretation of technical functions in terms of actual physical capacities or actual physical behaviour turns out to be a viable one, then it follows that the engineer's structure-function conception of technical artefacts fits nicely into the physical conceptual framework. If technical artefacts are physical structures with technical functions, and technical functions refer to actual physical input-output relations or physical capacities, then technical artefacts can be completely described in physical terms. Notions from the intentional conceptual framework are not necessary for characterizing technical artefacts, since all relevant technical properties of a technical artefact may be expressed within the physical conceptual framework. On this interpretation of the notion of function, technical artefacts are simply physical objects.

However, any interpretation of technical functions that tightly couples technical functions to actual physical capacities or actual behaviour by assuming that an object *x* has technical function *Y* if and only if it has certain physical capacities or shows certain behaviour, will run into serious problems. To show this we first have to bring to the surface an assumption we have made so far implicitly, namely that *the technical function of an object determines, in the sense of being a necessary and sufficient condition, the technical kind to which it belongs*. Roughly, technical kinds may be taken to correspond to the functional categories or classes we use in daily life to classify technical artefacts (see also the above quote from Ullman). So, a technical artefact is an instance of the technical kind telephone if it has the function of making telephone calls and a technical artefact that does not have the function to make telephone calls is not an instance of the technical kind telephone. This may seem an innocent assumption, but it is far from that, as we will see in due course.

If we stick to the assumption that functions determine technical kinds, then the interpretation of functions in terms of actual physical capacities runs into difficulties. This may be illustrated in the following way. Suppose that my telephone does not work because of a loose contact; so it does not have the right physical capacity and therefore not the function of making telephone calls.

[20] Note that Cummins' approach does not allow the attribution of functions to systems as a whole; functions are attributed to components of a system relative to a capacity of that system as a whole (and relative to an analytical account). This may not be a drawback for a function theory in the context of biology (where it is uncommon to attribute functions to organisms as a whole), but is surely is a serious shortcoming for a theory of technical functions.

[21] Cummins' aim is a non-teleological analysis of functions ascriptions, and therefore he avoids an appeal to human intentions as much as possible (1975, p. 751). It may be questioned, however, whether human intentions play no role in Cummins' interpretation of functions. Function attributions to components of systems are always relative to an analytical account. The content of the analytical account may not contain reference to human intentions, but the analytical account as such does. So after all Cummins' interpretation appears to be based, implicitly, on human intentions.

Nevertheless we still consider it to be a telephone (an instance of the technical kind telephone), albeit a malfunctioning one - I take it that it is needless to add that this line of reasoning does not apply to a telephone that has been smashed to pieces, since then there is no longer a telephone to malfunction in the first place. However, on the interpretation of functions as actual physical capacities, the telephone with the loose wire is not a telephone at all, since it does not have the physical capacity and therefore the function to make telephone calls. If functions determine technical kinds, and if we stick to the idea that a malfunctioning telephone is still a telephone, then having the appropriate physical structure cannot be a *necessary* condition for having the corresponding function. So, interpretations of technical functions along the lines presented above are unable to deal with malfunctioning technical artefacts. More generally these interpretations have difficulties in accounting for normative or evaluative claims about technical artefacts (such as "This printer ought to work." and "This is a good printer."). After all, this is not surprising since these approaches treat technical artefacts as similar to physical objects and normative or evaluative claims about physical objects ("This is a good electron.") in general make no sense.

Interpretations of technical functions in terms of physical capacities show still another shortcoming, since having the appropriate physical capacity can not also be *sufficient* for having a technical function, again on condition that functions determine technical kinds. Consider the situation in which I successfully use a coin for fastening or loosening a screw. So the coin has the appropriate physical capacity and may therefore be said to have the function of driving screws. That being so, the conclusion inevitably follows that the coin is a screwdriver, since by assumption function determines technical kinds. My successful use of the coin for driving screws, however, does not make it a screwdriver. It may make sense to say that I use the coin *as* a screwdriver, but surely it *is* not an object of that kind. The upshot of the foregoing appears to be that, given the assumption about functions and technical kinds, having the appropriate physical structure cannot be taken to be a necessary nor a sufficient condition for an object to have the corresponding function.

A problem that interpretations of functions along the lines sketched above run into is that they cannot deal with the distinction between *accidental* and *proper* functions of technical artefacts. The physical structure of a technical artefact has many physical capacities that make it possible to use that artefact for various purposes. Someone totally unfamiliar with nail clippers will have a hard time to find out which of the many input-output relations, that is, which physical capacity of the object depicted in Fig. 2.4 is its proper function and therefore to determine its technical kind. Interpretations of functions in terms of actual physical capacities lack the conceptual power to pick out which one is the intended or preferred capacity as its proper function; this proper function, it seems, determines to what technical kind the technical artefact belongs.

Various strategies may be followed to rescue the interpretation of functions in terms of actual physical capacities. One is to give up the assumption about functions determining technical kinds. Another one is to deny that a malfunctioning telephone, even if the cause of failure is only a loose wire, is still a telephone (more or less analogously to a false banknote not being a real banknote, or a fake gun not being a real gun).[22] A rather obvious way to try to amend this function interpretation is by coupling function to intended capacities or intended behaviour. So the malfunctioning telephone might still be a telephone because the physical structure involved is intended to have the capacity to make telephone calls, and a coin is not a screwdriver because it is not intended to have the physical capacity to drive screws (irrespective of whether it may actually be used in that way). That brings human intentions into the analysis.

2.5.2 *Functions and human intentions*

I now turn to an examination of attempts to interpret technical functions primarily or solely in terms of human intentions. Take the following analysis by Searle (1995) of what makes an object a screwdriver:

> It is, for example, an intrinsic feature of the object in front of me that it has a certain mass and a certain chemical composition. It is partly made of wood, the cells of which are composed of cellulose fibers, and also partly of metal, which is itself composed of metal alloy molecules. All these features are intrinsic. But it is also true to say of the very same object that it is a screwdriver. When I describe it as a screwdriver, I am specifying a feature of the object that is observer or user relative. It is a screwdriver only because people use it as (or made it for the purpose of, or regard it as) a screwdriver.

According to this view, what makes the object a screwdriver is not the sum total of its physical properties, its structure. Searle claims that the object has a function only in relation to human intentionality, which he characterizes broadly as the human capacity to represent objects and states of affairs in the world. Functions of technical artefacts only exist in the realm of human representations; human beings *assign* functions to objects (Searle 1995, p. 13 ff). It is a mistake to think that technical artefacts have functions independent of human intentions. So the object on my desk is a telephone because of people's intentions with regard to it: they think of it, use it, sell it, maintain it, and so on, as a telephone. Its function is an observer or mind-dependent feature.

Searle's main reason for relegating technical functions to the domain of human intentionality is their teleological nature. He claims that functions in general, not only technical functions but also biological and social functions, are assigned

---

[22] Duzi, Jespersen *et al.* (2010, p. 406 ff) present a formal analysis of two different accounts of malfunction, namely *subsective* versus *privative* malfunction; a subsective malfunctioning X is still an X, but a privative malfunctioning X is no longer an X; see also Jespersen and Carrara (2011). Whether in the case of technical artefacts we are dealing with subsective or privative forms of malfunction depends, of course, heavily on the cause of its malfunctioning; if a telephone malfunctions because of a loose wire it is most plausible to opt for the subsective form of malfunction; however, if it is smashed to pieces, then the privative form of malfunction appears to apply. Clearly, a sharp distinction between both forms of malfunctioning will be difficult to make; see Kroes (2003).

relative to the ends (interests or values) of human beings (Searle 1995, p. 15). In his view, it makes no sense to speak of functions of objects without reference to human ends. Technical artefacts are assigned functions because they are means for realizing human ends and this makes the notion of technical function a thoroughly teleological notion. I have expressed this teleological nature by saying that technical artefacts have a purpose, function, or a certain "*for-ness*": they are means *for* doing something or *for* achieving ends. Searle claims that any teleological feature of technical artefacts is always derivative of human ends.

Searle is certainly not the only one to defend the dependence of functions of technical artefacts on human intentions. McLaughlin (2001), to give another example, takes a similar position. He too claims that (2001, p. 60):

> [t]he function of an artefact is derivative from the purpose of some agent in making or appropriating the object; it is conferred on the object by the desires and beliefs of an agent. No agent, no purpose, no function.

The functions of technical artefacts may be changed simply by a change of mind (McLaughlin 2001, p. 52). The beliefs on the basis of which an agent attributes a function to an object even need not be true. So there is no guarantee that the attribution of a function to an object implies that that object actually has the capacity to realize that function. Function attribution to technical artefacts, according to McLaughlin, is not constrained by success criteria. If someone designs a new prototype of a can opener and it turns out to be a complete failure, the prototype can be said to have the function of opening cans (2001, p. 61).[23] The attribution of functions to artefacts is arbitrary in the sense that it is based on intentions of agents and these are completely external to the artefact.[24] Only in the case of the attribution of functions to components of more complex technical artefacts, there are according to McLaughlin constraints on the arbitrariness of attributing functions to those components, because these components causally contribute to capacities of the artefact as a whole of which they are part.

What is, at first sight, so puzzling about these intentional approaches to the functions of technical artefacts is that, if we cling to our assumption that the function assigned or attributed to an object determines its technical kind, the physical structure of technical artefacts does not seem to matter much in determining what kind of technical artefact we are dealing with. Only human intentions play a role. The two objects on my desk, one called 'stapler' and the other 'telephone', are examples of

---

[23] Millikan (1993, p. 22) makes a similar claim.

[24] McLaughlin remarks that the function of simple and even more complex technical artefacts can be arbitrarily changed, but that for more complex technical artefacts, in particular for its parts, it becomes (2001, p. 53) "somewhat more difficult for the armchair philosopher to think up plausible stories in which the function changes without any physical change". But why would a change in function attribution imply a change in physical structure if function attribution is in no way constrained by success, as he repeatedly stresses? If the newly designed can opener, that is a total failure, has the function of opening cans, then I see no reason why the stapler, used as a telephone, would not have the function of making telephone calls.

different kinds of technical objects. What makes the one a stapler and the other a telephone are, according to this line of thought, human intentions with regard to those objects. For something to have a particular function to φ in the sense of being assigned or attributed that function and therefore to be an artefact of the φ-kind, that is, to be an φ-er, humans must have the appropriate intentions: to design it as, use it as, buy it as an φ-er and so on. According to Searle (1995, p. 13) this implies that seeming to be a screwdriver is logically prior to being a screwdriver. The object on my desk is a stapler because it is assigned the function of a stapler and this assignment is based on humans treating it as a stapler etcetera. Similarly, McLaughlin claims that being an object of a functional (technical) kind is a status humans confer onto objects and these functional kinds are culturally determined (2001, p. 44). Human intentions are crucial in determining the technical kind of an object, not its physical structure. The latter plays only a derivative role in so far it may be the object of human intentions (for instance, someone may confer a specific function on an object on the basis of the belief that it has the appropriate physical structure).

Intentional approaches become even more puzzling when function attribution becomes as liberal as proposed by McLaughlin.[25] His account of function attribution is extremely permissive since there are no success constraints on function attribution to artefacts as a whole: whether or not the object is actually able to perform the function is not important for the function attribution. I have already observed that making success a *necessary* condition for function attribution is problematic; the malfunctioning telephone is still a telephone and still has the function of making telephone calls. But how can this case of genuine malfunctioning be distinguished from the unsuccessful use of the stapler as a telephone? It seems hardly reasonable to interpret this as a case of malfunctioning, that is, to call the stapler a malfunctioning telephone. Nevertheless, this is what McLaughlin's view appears to imply, if indeed the function attributed to the object determines its technical kind. Within his view it is not possible to distinguish between the two cases. He cannot make an appeal, for instance, to the successful use of other telephones of the same type in other situations for explaining the difference. That would be incompatible with his claim that function attribution does not depend on success.

The most obvious way to try to remedy these shortcomings is to let the physical structure play a more prominent role in attributing functions to technical artefacts in intentional approaches. This could be done by bringing in criteria of success with regard to the performance of technical functions. Function attribution may be restricted to cases in which the object involved has the appropriate physical capacities for bringing about the intended effects. But this is not an easy matter, since, as I already observed, success cannot be a necessary nor sufficient condition for function attribution - still always assuming that function attribution also determines technical kind. There appears to be no straightforward way of incorporating success

[25] As we will see later on in more detail (chapter 3), Searle's view on functions appears to be not as permissive as McLaughlin's since he interprets technical functions as causal agentive functions, which means that technical functions are performed on the basis of physical capacities.

criteria in intentional approaches to function attribution. The criteria should be flexible enough to justify function attribution to the malfunctioning telephone, and strict enough to exclude attributing the function of a telephone to a stapler.

The main lesson to be drawn from this admittedly rough and simplified sketch of two story lines about functions, one that ties functions primarily or exclusively to physical capacities, the other to human intentions, is that we need a theory of functions that ties technical functions to both physical structures and human intentions. In other words, technical functions, and therefore also technical artefacts, have a 'dual nature' for they are related to physical structures and human intentions.

2.6 The dual nature of technical artefacts

According to the view I will outline below and work out in more detail in the following two chapters technical functions are related to physical structures on the one hand and human intentions on the other. Because of this peculiar nature of technical functions, technical artefacts are different from physical objects and social objects. They have a hybrid, dual nature in the sense that it takes a combination of concepts from the physical and intentional conceptual frameworks to characterize technical artefacts (see Fig. 2.6).[26]

Herbert Simon's theory about artificial things as exposed in his classic *The sciences of the artificial* (1996 (1969)) is a nice stepping stone for introducing the dual nature view. For Simon the artificial sciences, including engineering, are about adapting the world to human needs and desires. One way of doing this is by

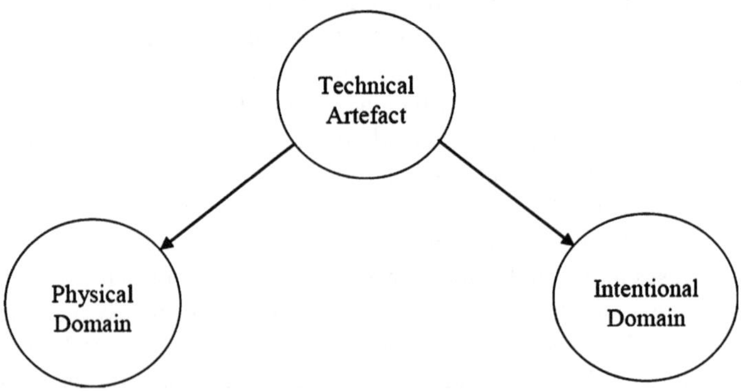

Fig. 2.6 The dual nature of technical artefacts

[26] In this figure and the following ones the solid arrows stand for conceptual implication: the notion of a technical artefact conceptually implies the notion of a domain of physical objects and a domain of human intentions.

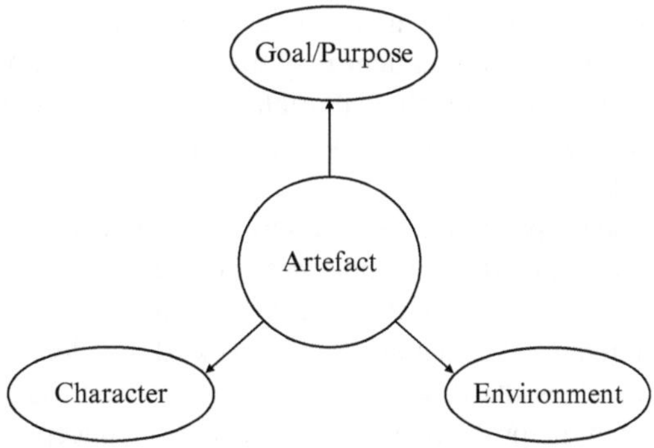

Fig. 2.7 Simon's analysis of a technical artefact

creating technical artefacts, things synthesized by humans for a certain purpose. The functional or purposeful nature of technical artefacts is analysed by Simon in the following way (*ibidem*):

> Let us look a little more closely at the functional or purposeful aspect of artificial things. Fulfillment of purpose or adaptation to a goal involves a relation among three terms: the purpose or goal, the character of the artifact, and the environment in which the artifact performs.

For instance, the purpose of a clock is to tell time and the character of the clock refers to its physical makeup (gears, springs etc. for a mechanical clock). Finally, the environment is important because not every kind of clock is useful in every environment; sun dials can only perform their function in sunny climates. Simon's analysis of artefacts is represented in a schematic way in Fig. 2.7.

Simon distinguishes between the inner and outer environment of a technical artefact, which corresponds to two different ways of looking at technical artefacts. Looked at from the outer environment, a technical artefact presents itself primarily as something, whatever its inner environment, that fulfils a certain goal, purpose or function. From this perspective the artefact is characterised primarily in a functional way; the inner environment, that which realizes the function, remains a black box. Looked at from the inner environment, the artefact is described as some kind of physical (biological) system; from this perspective, the goal that it fulfils in the environment remains a black box. As Simon (1996 (1969), p. 7) remarks "*Given* an airplane, or *given* a bird, we can analyze them by the methods of natural science without any particular attention to purpose or adaptation, without reference to the interface between what I have called the inner and outer environments." Simon's distinction between the outer and inner environment view of technical artefacts runs closely parallel to Chandrasekaran and Josephson's distinction between the environment-centric and device-centric views on function (Chandrasekaran and Josephson 2000).

These two different ways of characterising artefacts, from the perspective of their inner and outer environment, lead us back to our main question of how technical artefacts fit into the physical and intentional conceptualizations of the world. This becomes clear as soon as we take a look at the conceptual frameworks used to describe the inner and outer environments. For technical artefacts, the description of the inner environment may make use of the physical conceptualization; as Simon states, once a technical artefact is at hand, we can analyse what is in the technical artefact by the methods of the natural sciences; no appeal to purpose, functions or adaptation is necessary. The outer environment is described partly in terms of goals and purposes, that is, is described partly in a teleological way. If we are dealing with technical artefacts, then these goals and purposes are related to human ends and so the description of the outer environment will have to make use of concepts from the intentional conceptual framework. If an adequate description of technical artefacts involves the description of elements from both the inner and outer environment, then the problem arises of how to use concepts from both frameworks side by side in a coherent way in the description of technical artefacts.[27]

My starting point for exploring this issue is the engineer's structure-function conception of technical artefacts. I take technical artefacts to be physical structures with technical functions. To that I add the conclusion drawn from our discussion of the two story lines about functions, namely that an adequate interpretation of technical functions will have to ground technical functions in physical as well as intentional features of the world (I will return to the issue of the adequacy conditions for theories of functions in chapter 4). This makes a technical artefact a hybrid kind of object that does not fit in either the physical or the intentional conceptualisation. Looked upon as merely physical objects, technical artefacts fit into the physical conceptualisation of the world; what the artefact does can be explained in terms of causal processes. But as a mere physical object, it is not a technical artefact. Without its function, the object loses its status as a technical artefact. This means that technical artefacts cannot be described exhaustively within the physical conceptualisation, since it has no place for its functional features. But neither can it be described exhaustively within the intentional conceptualisation. The latter has no problem with describing the function of a technical artefact in terms of what it is for, and to relate this function to human ends. But a means-end description of a technical artefact effectively black boxes its physical structure. As I pointed out earlier, such a description has the following structure: x is for φ-ing, or x is a means to φ-ing, where x stands for some physical object and φ-ing for some activity. Of course, the description of φ-ing, of what it is for, will contain physical concepts, but that does not provide a more detailed description of x beyond the fact that x is some physical object. In a functional description of an object x, reference to physical concepts occurs after the "for"-operator; the object x itself remains a black box. The intentional conceptualisation has no way of dealing with the physical features of a technical artefact itself; it leaves open how a description of the functional features of a technical artefact coheres with its physical features. Hence the conclusion that technical artefacts have a dual nature: on the one hand they are physical, on the other intentional objects (Kroes and Meijers 2006; Kroes 2010).

[27] A much more daunting problem would be how to integrate the physical and intentional conceptual frameworks into one coherent conceptualization in general. That problem falls outside the scope of this book.

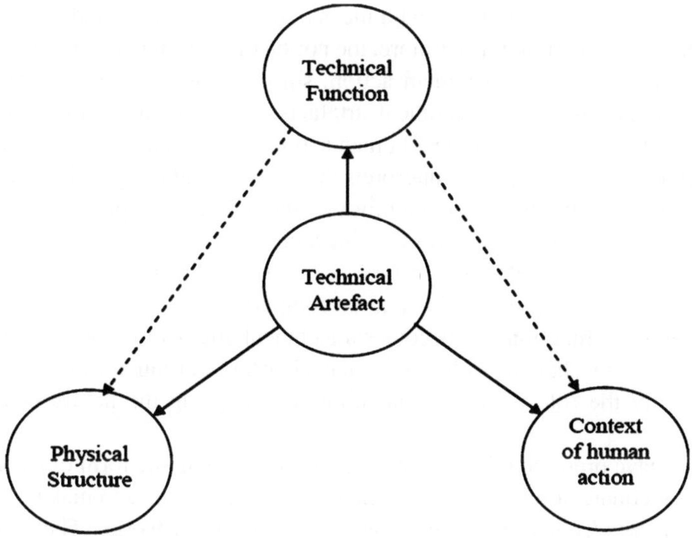

Fig. 2.8 A conceptual anatomy of the concept of technical artefact

According to the above line of thought, a conceptual analysis of the notion of technical artefact must refer to three key notions, namely the notion of a physical structure, of a technical function and of a context of (intentional) human action (see Fig. 2.8; solid arrows).[28] The notion of a technical function, moreover, is related to the notion of a physical structure and of (intentional) human action (indicated by the broken arrows).

Various kinds of contexts of intentional human action may be distinguished, two of which are of particular interest to us, namely the one in which technical artefacts are designed (and produced) and the one in which they are used. The distinction between these two kinds of contexts will play a major role later on in our analysis of the notion of function and of technical kinds.

There are some notable differences between our analysis of technical artefacts and Simon's. Simon's notion of goal or purpose has been replaced by the notion of function. I assume that technical artefacts have functions but not goals (in the sense of an aim or an end (telos)).[29] That notion has its place in a context of intentional

[28] In a more or less similar way, Losonsky (1990, p. 84) analyses the nature of artefacts in terms of the following three features: internal structure, purpose and manner of use.

[29] This assumption may be questioned on the grounds that nowadays engineers design complex adaptive systems that appear to have goals of their own (for instance, missiles that track their targets). Even if it is granted that these technical systems have genuine goals in the sense that these goals are somehow independent of the intentions of designers and/or users, such goals have to be distinguished from goals (ends) that play role in contexts of human actions. Whereas these human goals are related to intentional states, it is highly questionable that this is also the case for the goals of complex adaptive systems. In order to avoid confusion between these two kinds of goals, I prefer to refer to goals only in a context of human action.

human action; within such a context a means used to achieve a goal (end, aim) has or is attributed a function. Furthermore, the notion of environment has been replaced by the notion of context of human action, since not any kind of environment is relevant for the analysis of technical artefacts; only references to environments comprising a context of intentional human action are appropriate. In his example of the sun dial, for instance, Simon interprets the environment in a physical way (sunny climates are the required environment for sun dials). But this is problematic. It is not this physical environment that turns the object involved, a stick which casts a shadow on a surface, into an artefact of the type sun dial. Only within a context of human action (e.g. of ordering events in time or comparing time intervals) this physical object acquires a function and becomes a technical artefact (a time keeping device or clock). So, our changes to Simon's analysis bring out much more prominently and explicitly the role of human intentions in analysing the nature of technical artefacts.

So far I have presented the dual-nature thesis mainly in the form of a conceptual claim: an adequate description of a technical artefact will have to make use of concepts from the physical and intentional conceptual frameworks. The dual-nature view, however, may also be interpreted in an epistemological and ontological way. From an epistemological point of view the dual-nature thesis states that knowledge about technical artefact falls apart in two distinct kinds. We have already briefly touched upon the epistemological interpretation when claiming that engineers make use of two different kinds of knowledge about technical artefacts, namely, knowledge of their structural and of their functional features. From an ontological point of view the dual-nature thesis may be interpreted as stating that technical artefacts are somehow constituted by physical objects and by human intentions. However the constitution relation may be interpreted, this means that technical artefacts are ontologically mind-dependent entities and as such are ontologically different from physical objects. An object cannot be a technical artefact by virtue of its physical properties alone. Being a technical artefact also involves functional properties and these in turn involve human intentions (in section 3.3 I will discuss in detail epistemic and ontological theories of technical functions).

These various interpretations of the dual-nature thesis raise questions about how they are related. It may be questioned whether the conceptual and epistemological interpretations are really different, since the conceptual interpretation is based on the notions used in the most accurate descriptions available. Apart from that, there is the issue about how the epistemological and the ontological interpretations are or are to be related. I will mainly sidestep these questions for they would lead into general philosophical debates about (the possibility of) conceptual analysis and about how epistemology and ontology are related. In section 3.3, though, I will briefly enter into a discussion about how I see the connection between epistemic and ontological theories of functions. For the moment, one important remark about the ontological interpretation has to be added. I assume throughout this book a basic ontology figuring both physical objects and human intentions. This basic ontology may be taken to be a combination of the ontologies that may be associated with the two basic conceptual frameworks. My claim that technical artefacts are constituted by physical objects and human intentions is to be taken relative to this basic ontology. Of course, this basic ontology raises many questions, in particular about how in general human intentions are ontologically related to (features of) physical objects; these questions, however, fall squarely outside the scope of this book. The view on technical artefacts elaborated here may be said to be robust against solutions of these issues in the sense that both physical objects (states) and human intentions, however they may be ontologically construed in relation to each other, are constitutive for an object to be a technical artefact.

2.7 Conclusion

The differences that we have encountered so far between technical artefacts, physical objects and social objects may be summarized in the following way. With regard to technical artefacts, the following three questions in principle always make sense:

a) What is this object for?
b) What is this object made of?
c) How is this object to be used?

The first question concerns the function of the technical artefact, the second its physical structure, and the third its manual or use-plan. Here we recognize the three elements of technical artefacts represented in Fig. 2.8. For physical objects only the second question makes sense; the first and third questions are meaningless: physical objects have no functions and they do not come with a manual. For many social objects the first and third questions do make sense; they have a function and the 'manual' describes what actions are required or what rules are to be followed for the social object to perform its function. For social objects it is not so clear how the second question is to be interpreted. Surely, social objects are not made of material stuff, although physical objects (phenomena) may be involved, as in the case of money.

I have argued that technical artefacts have a dual nature because their functional features are on the one hand related to their physical features and on the other to human intentions. An adequate description of technical artefacts always is based upon a combination of concepts from the physical and intentional conceptual frameworks. In the course of our exploration of the nature of technical artefacts we have come across a number of issues that need further clarification. The most important are:

- How are malfunction claims, more generally evaluative and normative claims about technical artefacts to be interpreted?
- What does it mean to say that an object is an instance of a technical kind and does the function of a technical artefact determine its technical kind, as we have assumed so far?

References

Chandrasekaran, Balakrishnan. 2005. Representing function: Relating functional representation and functional modelling streams. *Artificial Intelligence for Engineering Design, Analysis, and Manufacturing* 19:65–74.

Chandrasekaran, Balakrishnan, and John R. Josephson. 2000. Function in device representation. *Engineering with Computers* 16 (3–4):162–177.

Chittaro, Luca, and Amruth N. Kumar. 1998. Reasoning about function and its applications to engineering. *Artificial intelligence in engineering* 12:331–336.

Cummins, Robert. 1975. Functional analysis. *Journal of Philosophy* 72 (20):741–765.

Davidson, Donald. 1963. Actions, reasons and causes. *Journal of Philosophy* 60:685–700.

Dennett, Daniel C. 1987. *The intentional stance*. Cambridge, Massachusetts: The MIT Press.

Dipert, Randall R. 1993. *Artifacts, Art Works, and Agency*. Philadelphia: Temple University Press.

Dretske, Fred I. 2006. Minimal rationality. In *Rational animals?*, eds. Susan L. Hurley, and Matthew Nudds. Oxford: Oxford University Press.

Duzi, Maire, Bjorn Jespersen, and Pavel Materna. 2010. *Procedural semantics for hyperintensional logic; foundations and applications of transparent intensional logic*. Springer.

Franssen, M. 2008. Design, use, and the physical and intentional aspects of technical artifacts. In *Philosophy and design; from engineering to architecture*, eds. Pieter Vermaas, Peter Kroes, Andrew Light, and Steven A. Moore, 21–35. Springer.

Hacking, Ian. 1983. *Representing and intervening; introductory topics in the philosophy of natural science*. Cambridge: Cambridge University Press.

Hilpinen, Risto. 2004. Artifact. In *The Stanford Encyclopedia of Philosophy (Fall 2004 Edition)*, ed. Edward N. Zalta. (editor): http://plato.stanford.edu/archives/fall2004/entries/artifact/.

Jespersen, Bjorn, and Massimiliano Carrara. 2011. Two conceptions of technical malfunction. *Theoria* 77:117–138.

Koepsell, David. 2009. *Who owns you? The corporate gold rush to patent your genes*. Oxford: Wiley-Blackwell.

Kroes, P. 2003. Screwdriver philosophy; Searle's analysis of technical functions. *Techné* 6 (3):22–35.

Kroes, P., and A. Meijers. 2006. The dual nature of technical artefacts - Introduction. *Studies in History and Philosophy of Science* 37 (1):1–4.

Kroes, Peter. 2003. Physics, experiments and the concept of nature. In *The philosophy of scientific experimentation*, ed. H. Radder, 68–86. Pittsburgh: University of Pittsburgh Press.

Kroes, Peter. 2006. Coherence of structural and functional descriptions of technical artefacts. *Studies in History and Philosophy of Science* 37 (1):137–151.

Kroes, Peter. 2010. Engineering and the dual nature of technical artefacts. *Cambridge journal of economics* 34:51–62.

Kroes, Peter, Maarten Franssen, Ibo van de Poel, and Maarten Ottens. 2006. Treating socio-technical systems as engineering systems: some conceptual problems. *Systems research and behavioral science* 23:803–814.

Krohs, Ulrich, and Peter Kroes. 2009. Philosophical perspectives on organismic and artifactual functions. In *Functions in biological and artificial worlds; comparative philosophical perspectives*, eds. Ulrich Krohs, and Peter Kroes, 3–12. Cambridge, Mass.: The MIT Press.

Longy, Françoise. 2009. How biological, cultural, and intended functions combine. In *Functions in biological and artificial worlds; comparative philosophical perspectives*, eds. Ulrich Krohs, and Peter Kroes, 51–67. Cambridge Mass.: MIT.

Losonsky, Michael. 1990. The nature of artifacts. *Philosophy* 65:81–88.

McLaughlin, Peter. 2001. *What Functions Explain: Functional Explanation and Self-Reproducing Systems*. Cambridge: Cambridge University Press.

Millikan, Ruth. 1993. *White Queen Psychology and Other Essays for Alice*. Cambridge, MA: MIT Press.

Otto, Kevin N., and Kristin L. Wood. 2001. *Product design; techniques in reverse engineering and new product development*. Upper Sadle River (NJ): Prentice Hall.

Pahl, G., and W. Beitz. 1996. *Engineering Design: A Systematic Approach.* London: Springer Verlag.

Perlman, Mark. 2009. Changing the mission of theories of teleology: DOs and DON'Ts for thinking about function. In *Functions in biological and artificial worlds; comparative philosophical perspectives*, eds. Ulrich Krohs, and Peter Kroes, 17–36. Cambridge Mass.: MIT Press.

Preston, Beth. 1998. Why is a wing like a spoon? A pluralist theory of function. *Journal of Philosophy* 95 (5):215–254.

Preston, Beth. 2008. The shrinkage factor: comment on Lynne Rudder Baker's "The shrinking difference between artifacts and natural objects". *APA newsletter on philosophy and computers* 8 (1):26–28.

Roozenburg, N.F.M., and J. Eekels. 1995. *Product Design: Fundamentals and Methods.* New York: John Wiley & Sons.

Rosenman, M.A., and J.S. Gero. 1994. The what, the how, and the why in design. *Applied artificial intelligence* 8:199 - 218.

Searle, John. 1995. *The Construction of Social Reality.* London: Penguin Books.

Simon, Herbert A. 1996 (1969). *The Sciences of the Artificial.* 3 rd ed. Aufl. Cambridge, MA: MIT Press.

Sperber, Dan. 2007. Seedless grapes: nature and culture. In *Creations of the mind: therories of artifacts and their representation*, eds. Eric Margolis, and Stephen Laurence, 124–137. Oxford: Oxford University Press.

Thomasson, Amie L. 2007. Artifacts and human concepts. In *Creations of the mind: essays on artifacts and their representations*, eds. Stephen Laurence, and Eric Margolis, 52–73. Oxford: Oxford University Press.

Ullman, David G. 1992. *The Mechanical Design Process.* New York: McGraw-Hill.

Vermaas, Pieter E., and Kees Dorst. 2007. On the conceptual framework of John Gero's FBS-model and the prescriptive aims of design methodology. *Design Studies* 28 (2):133 - 157.

Wittgenstein, Ludwig, and G. E. M. Anscombe. 2003. *Philosophical investigations: the German text, with a revised English translation.* 3 rd Aufl. Malden, MA,: Blackwell Pub.

[illegible]

Chapter 3
Theories of technical functions

3.1 Teleology and function theories

As is well-known Aristotle's view on nature, including inanimate natural objects, was thoroughly teleological. He distinguishes between things that exist by nature and those that exist by other causes (*Physica*, book II, 192^b). Things that exist by nature, natural things, are characterised by the fact that they carry within themselves their principle of change. Nature, according to Aristotle, is "a source or cause of being moved and of being at rest" within a thing, which it has by virtue of being that thing. In other words, a natural thing is a thing that, by being that thing, carries within itself its own principle of motion or change (for biological organisms this includes change in the sense of reproduction). This principle of motion is called its nature and is directed towards the realisation of its goal or telos.[1] For example, a seed of a beech tree, by virtue of being a beech seed, carries within itself its principle of growth into a beech tree. The growth of a beech tree according to its intrinsic principle of motion is a natural phenomenon.

In contrast to its natural growth, the transformation of a beech tree into a bed is not a change that finds its origin in the beech tree qua beech tree. This is not a natural change (a change according to its nature), and the bed is not a natural, but an artificial object. Moreover, the bed, qua bed, does not carry within itself its own principle of motion.[2] The cause of this change lies outside the bed, namely in the artisan who

[1] Note that Aristotle defines nature as a cause or principle of motion and not as a collection of things (and/or phenomena). In line with Aristotle's definition of nature as a cause, nature in the sense of the natural world (a collection of things and/or phenomena) may be defined as those things which "exist by nature", i.e., which carry within themselves their own principle of motion.

[2] As Aristotle remarks (192^b): "…a bed and a coat and anything else of that sort, *qua* receiving these designations – i.e. in so far as they are products of art – have no innate impulse to change. But in so far as they happen to be composed of stone or of earth or of a mixture of the two, they *do* have such an impulse, and just to that extent – which seems to indicate that nature is a source or cause of being moved and being at rest in that to which it belongs primarily, in virtue of itself and not in virtue of a concomitant attribute." Thus (193^a) "if you planted a bed and the rotting wood acquired the power of sending up a shoot, it would not be a bed that would come up, but *wood*."

P. Kroes, *Technical Artefacts: Creations of Mind and Matter: A Philosophy of Engineering Design*, Philosophy of Engineering and Technology 6,
DOI 10.1007/978-94-007-3940-6_3, © Springer Science+Business Media B.V. 2012

produces the bed for a certain purpose. The bed contrary to the beech tree has no intrinsic principle of motion, no intrinsic telos which it tries to realize. As a technical artefact it has literally no nature in the Aristotelian sense. If the bed, in modern terms, is said to have a function it has that function only in relation to some human end and not in relation to an end of itself. This lack of a nature of its own is the reason why the bed for performing its function will in time need maintenance by the artisan since natural processes occurring in the wood may lead to its decay and this decay may interfere with the fulfilment of its function. Objects that exist by nature stand not in need of maintenance; on the contrary, any human interference with such objects, including maintenance, constitutes a disturbance of their nature.[3]

According to Aristotle not only biological phenomena are teleological in nature, but also physical phenomena. For instance, the free fall of a heavy body toward the centre of the Earth is the result of the striving of that heavy body to realize its intrinsic end. That end is to be located at its natural place which is at the centre of the universe which in Aristotle's picture of the world coincides with the centre of the earth.

Over the course of history Aristotle's teleological conception of nature has been given up, first for the physical world during the Scientific Revolution of the sixteenth- and seventeenth centuries and for the biological world beginning with the rise of evolution theory in the second half of the nineteenth century. In the present day conception of the physical world teleological notions play no role of any significance, if at all.[4] As far as the description of the physical world is concerned there is no place for ends, goals, final causes or *telos* of any kind. Nevertheless, from a modern perspective a case could be made, within an Aristotelian line of thought, that physical objects carry their principles of change within themselves if we take the laws of nature to be these principles of change (Kroes and Vermaas 2008).[5] These principles, however, are not related to or presuppose ends of or in physical objects. Just as technical artefacts have no intrinsic ends for Aristotle, physical objects have no ends for the modern physicist.

Within the biological sciences, however, attempts to purge the conceptual framework used in describing biological phenomena and organisms from teleological elements turns out to be much more problematic. The difficulty mainly resides in the use of the notion of function. This use is virtually inevitable, for instance in anatomy. That being the case, the most obvious way to avoid teleological notions in the description of the biological world is to develop a naturalistic notion of function that avoids any reference to teleological concepts such as purposes and goals.

[3] This idea still lies at the bottom of the modern conception of nature as that which is untouched by human beings.

[4] According to Perlman (2004, p. 4) by "the twentieth century, analytic philosophers were positively allergic to any mention of teleology or teleological function. It was seen as an insidious metaphysical notion that was to be tossed out with the rest of metaphysics."

[5] According to this line of reasoning, physical objects may be said to have a nature and to behave natural; this would be true irrespective of whether these physical objects are human-made or not; see the discussion in section 2.3.

The recent revival of interest in function theories is indeed mainly driven by a desire to naturalise the notion of function and has resulted in various theories for functions in general and for biological functions in particular. The task of classifying these theories is not an easy one.[6] Here we will just mention the kinds of theories that are most relevant as a background for our discussion of technical functions.

Following Perlman the category of naturalistic theories of functions may be divided up into two sub-categories, one sub-category containing theories that consider functions to be reducible to more basic natural phenomena, and one that considers functions to be natural, but irreducible.[7] Searle's theory of functions, that we have come across already briefly in the previous chapter and which will be discussed in more detail below, belongs to the latter category of naturalistic but non-reductionist theories. The category of reductionist naturalistic theories is made up of three different kinds of theories, depending on whether functions are reduced to actual, past or future natural phenomena. Cummins theory, that interprets functions in terms of causal roles and that we already briefly touched upon in chapter 2, is an example of the first kind. Evolutionary (etiological) theories of function exemplify the second kind. Roughly, they interpret the function of a biological trait as that effect for which the trait was selected in its lineage of ancestors. These theories reduce functions to past phenomena. The third kind contains theories that interpret functions in terms of future phenomena. Within the biological domain these theories couple functions to survival enhancing propensities of organisms and to their fitness; functions are construed as dispositions which makes them forward-looking. More recently also technical functions have received attention within discussions about these theories. One of the bones of contention is whether it will be possible to come up with a unified theory of functions that will be able to account for biological and technical functions, or whether a pluralist account is called for.[8] One argument for a pluralist account is that in contrast to biological functions, the teleological nature of technical functions is generally considered to be unproblematic since technical functions are taken to be intimately related to intentional human action of which the teleological nature is taken for granted.

In so far function theories for technical artefacts are concerned, the following classification has been proposed by Houkes and Vermaas (2003; 2010, Ch. 3). They distinguish three different kinds of function theories, called the intentional, causal-role and evolutionist theories of function. On the intentional (I-) function theories, the functions of technical artefacts are determined by the intentions, beliefs and actions of human agents. Agents ascribe functions to technical artefacts by embedding them in means-end relations and functional descriptions of technical artefacts explicitly or implicitly refer to goals of human agents. Searle's and McLaughlin's theories of

[6] For a general survey, see Perlman (2004)

[7] For more details and references, see Perlman (2004).

[8] See Preston (1998); for a critique Preston's position, see Millikan (1999); a defence of unitary theories of functions based on the notion of design may be found in Kitcher (1998) and Krohs (2004, 2009).

functions belong to this category. In causal-role (C-) theories of functions, exemplified by Cummins theory, the function of an object is related to the causal role it plays in a larger system of which it is a part. Function ascriptions to parts or components of the encompassing system are based on explanations of the capacities of that system in terms of dispositions of its components. Finally, evolutionary (E-) theories relate functions of technical artefacts to long-term reproduction and selection histories of their predecessors. Roughly, the function of an object is the effect for which its predecessors were selected. Preston's theory, to be discussed below, belongs to this category. As we will see later on Houkes and Vermaas use elements from these three kinds of function theories to construct their own ICE-theory of technical functions.

These classifications of various kinds of function theories do not run very much apart (see (Perlman 2004, p. 33). What is important to note is that there appear to be three different kinds of 'building blocks' out of which theories of technical functions are constructed, namely actual physical properties of technical artefacts, actual intentions of human agents (designers, users, others) and past or future events (such as past and future selection events). Before we can have a closer look at how these building blocks figure in various theories of function, it will be necessary to distinguish between two different kinds of functional properties of objects. An object may be related to a technical function by claiming

(1) that it has or may be attributed that function, or
(2) that it is an instance of the corresponding technical kind.

Theories of technical functions focus primarily on the first way of relating an object to a technical function, without paying much attention to what it means for an object to be an instance of a technical kind. As we have already pointed out in the previous chapter, they often assume implicitly that the attribution to an object of a function, as defined by those theories, amounts to the same as claiming that that object is an instance of the corresponding technical kind. That, however, is a problematic assumption that, I will argue, causes problems for theories of functions. What is needed to solve these problems is a clear insight into how the notion of function defined in theories of technical functions is related to the notion of function employed in theories of technical kinds. In other words, besides a theory of technical functions we also need a theory about technical kinds. Let us first, then, have a closer look at the two functional properties involved in all of this.

3.2 Functional properties

My focus is on technical artefacts whose functions are realized by physical objects or systems (so I am not considering the functions of processes or states of affairs, although I see no principled reason why the following analysis would not apply in those cases). Any such technical artefact may *prima facie* be characterized as a physical object, X, that on top of its physical properties has a functional property, to φ, when used in the appropriate way, for instance to drive screws. X is an element of the set of physical objects or systems, typically physical objects constructed by technicians or engineers and φ is an element of a particular set of verbs (to cut, to copy, to transport, to amplify etc.).[9]

The statement "*X* has the function to φ" I take to have the same meaning as the statement "The function of *X* is to φ".[10] The use of the definite article in the expression "the function" does not imply that to φ is *X*'s exclusive function nor that it is its proper function. The function to φ covers any kind of function (proper, accidental or otherwise). I interpret these statements as expressing the attribution of some kind of property, namely a *functional* property, to an object. I leave open the possibility that the property attributed may be a relational property, for instance relative to a specific context ("*X* has the function to φ relative to *R*" etc.). It is not a straightforward matter to express this attribution of a property in the traditional canonical form of "*S* is *P*" ("This apple is red."). One way to proceed is to assume that the statements "*X* has the function to φ" and "The function of *X* is to φ" have the same meaning as "*X* is for φ-ing". On the assumption that we may treat "for φ-ing" as a property, the attribution to *X* of the functional property to φ may be formulated canonically as "*X* is *P*" with *P* standing for "for φ-ing".[11] An object *X* may have several functional properties at the same time, so the statements "*X* is for φ-ing" and "*X* is for ψ-ing" may in principle be true simultaneously (think of multi-functional technical artefacts such as a Swiss army knife).

There is yet another kind of statements in which functions are coupled to objects, namely statements of the kind "*X* is a φ-er"; for instance the statement "*X* is a screwdriver". I take this statement to mean that *X* has a property *P*, namely of being a screwdriver, and having this property is tantamount to being an instance of the technical kind 'screwdriver'.[12] As before, technical kinds roughly correspond to the classes in which we categorize technical artefacts. In the following I will mainly consider forms of malfunctioning technical artefacts that allow for a subsective interpretation of malfunctioning (see (Duzi et al. 2010; Jespersen and Carrara 2011)). This means that, for instance, a malfunctioning screwdriver is still a screwdriver. On this condition, the class of all φ-ers may be subdivided into two subclasses, one of well functioning φ-ers and one of malfunctioning φ-ers.

---

[9] *X* may also stand for a natural object that performs a function, but it is debatable whether in that case we are dealing with technical *artefacts*, because the physical object involved is not a human-made construction.

[10] Strictly speaking we should say: "*X* has the function to φ with" and "The function of *X* is to φ with", which brings out more clearly the role of the human agent (be it the designer who intended the technical artefact to be used in a certain way or the user); to avoid cumbersome language I will stick to the present formulation.

[11] In the following I will use expressions like 'the property of *being* for φ-ing' as synonymous with 'the property 'for φ-ing'.

[12] Treating kinds and types as properties (universals) may raise problems; Wetzel (2006) notes that types (kinds) are not as obviously predicable as classic examples of universals (properties) because they are often referred to by singular terms (e.g., the ivory-billed woodpecker). I assume that such problems do not arise with regard to treating being an instance of a technical kind as properties.

Now let us see how the classes of all objects that are for φ-ing and of all objects that are φ-ers might be related to each other. I assume that the class of all objects that are for φ-ing can be divided schematically into two subclasses, one containing all objects that are capable of φ-ing and one all objects that are not. Furthermore I assume for the time being that *X* is a φ-er implies *X* is for φ-ing.[13] Thus, *X* is a screwdriver implies that *X* is for driving screws. This does not mean that *X* is actually used for driving screws; in a specific context the screwdriver may have the function of opening tin cans. Then, *X* has more than one function at the same time. In chapter 4, where I analyse the distinction between having a proper function and being a member of a technical kind in more detail, we will see that the assumption that *X* is a φ-er implies *X* is for φ-ing will have to be qualified. The reason is that an object may be a screwdriver, whereas its proper function, in a sense to be defined later on, is not to drive screws. Think of a screwdriver in a museum: it is a screwdriver, but its proper function is not to drive screws; the same applies for instance to Duchamp's *pissoir*.

In section 2.5.1 I already indicated that the reverse implication, from *X* has the function to φ (is for φ-ing) to *X* is an instance of the technical kind φ-er, is problematic. I leave open the possibility that the reverse is not true and thus that the meaning of the statements "*X* is for φ-ing" and "*X* is a φ-er" may be construed in such a way that they are not the same. An object may be for φ-ing without being a φ-er. Think of a coin that in a particular situation is being used as a screwdriver. In that context, the coin may be said to be for driving screws (for φ-ing), without being a screwdriver (a φ-er). So, the class of all objects with the property of being a φ-er is a proper subclass of all objects with the property 'for φ-ing'. The possible relations between the classes with various functional properties are represented succinctly in Fig. 3.1. Note that the subclass of all non φ-ers not capable of φ-ing may be a rather problematic one. Suppose that someone tries to loosen a screw with a nail but fails systematically because the nail lacks the appropriate physical capacities. It may be argued that because the nail is used for loosening a screw, it is attributed the functional property of loosening screws; so the nail is for loosening screws (for φ-ing) but not capable of it. However, since we are dealing here with a case of failed use it may also be argued that the attribution of the functional property is unjustified. Then, the nail is simply not for loosening screws (not for φ-ing). So, whether the subclass of all non φ-ers not capable of φ-ing is empty or not, depends on which theory of function attribution is adopted.

[13] This assumption may be questioned. Consider, for instance, a model boat. If it is taken to be a boat, then it certainly does not have the function of (is not for) transporting people or goods over water (see Bloom (1996) and Thomasson (2007)). Whether a model boat is a real boat, however, is a controversial claim. After all, when a model boat is taken to be a real boat and it is a model of a real (actual) boat, then it will be necessary to distinguish between different senses of what it means to be a real boat. I will interpret the predicate 'model' in such instances in a privative sense and therefore they cannot be taken to be counterexamples to the assumption that "*X* is a φ-er" implies "*X* is for φ-ing".

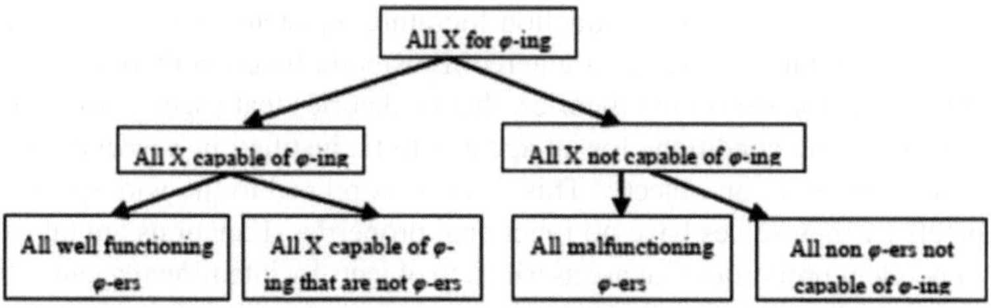

Fig. 3.1 Subdivision of the class of all X with the property 'for φ-ing'

To summarize, I distinguish between two kinds of statements in which functional properties are attributed to an object *X*, namely

(1) "*X* is for φ-ing", and
(2) "*X* is a φ-er'.

If *X* is a φ-er, then this implies that *X* is for φ-ing, but not the other way around. Finally, I take statements like "*X* functions as a φ-er' to be a special case of statements of the first type, that is, a special case of statements of the kind "*X* is for φ-ing"; in these statements the function associated with being a φ-er is attributed to *X*.

3.3 Epistemic and ontological theories of functions

Still one preparatory step is necessary before we can turn to reviewing theories of technical functions, namely an explication of the meaning of an object *having* the properties of being for φ-ing or being a φ-er. This may be done from an epistemic and an ontological point of view (Kroes 2010a, 2010b).[14] An epistemic explication focuses on what it means for an agent *A* to know or to have some form of justified belief that *X* has the functional properties of being for φ-ing or being a φ-er. Its aim is to define knowledge of functional properties of *X* in terms of knowledge of other kinds of properties of *X* (or more general epistemic conditions to be imposed on the agent *A*). This, of course, makes only sense if knowledge of functional properties is not considered to be some kind of basic or primitive knowledge itself. Ontological explications aim at defining functional properties in terms of what are considered to be more basic ontological properties.

Epistemic theories of functions have the following general form:

> Agent *A* knows that *X* is for φ-ing or is a φ-er iff agent *A* knows that *X* has such and such properties.

[14] For the distinction between epistemic and ontological theories of functions, see also Vermaas (2009).

However, going through the function literature, epistemic function theories of this form are seldom or never encountered. Epistemic function theories often take the form of function *ascription* theories, that is, theories that specify necessary and sufficient epistemic conditions for an agent *A* to be justified in *ascribing* a certain functional property to an object.[15] This, I think, is related to the widespread belief that objects by themselves have no functional properties. Functions are taken to be mind-dependent properties that are ascribed to objects by intentional agents. To stay in line with the literature I will also analyse epistemic function theories as function ascription theories without, however, thereby implying that functions are mind-dependent.

Given the two different kinds of functional properties, I will distinguish between two general types of epistemic function theories. The first type, to be called *theories of function ascription*, concerns the ascription of the functional property of being for φ-ing, the second, to be called *theories of function kind ascription*, concerns the functional property of being a φ-er. Ideally, epistemic theories of ascribing functional properties should state a set of conditions, each of which is necessary and together sufficient for an agent *A* to justifiably ascribe the properties of being for φ-ing and being a φ-er to an object *X*:

Epistemic theory of Function Ascription:

Agent *A* is justified in ascribing the functional property of being for φ-ing to object *X* iff agent *A* knows that *C1, C2,…,Cm*.

Epistemic theory of Function Kind Ascription:

Agent *A* is justified in ascribing the functional property of being a φ-er to object *X* iff agent *A* knows that *K1, K2,…, Kn*.

The set of conditions *K1…Kn* will have to include the set of conditions *C1…Cm* because, as we remarked above, function kind ascription implies function ascription, but not the other way around.

In order to avoid confusion, it is important to point out an ambiguity in the notion of function ascription. As Hansson (2006, p. 20–21) remarks the notion of function ascription is ambiguous between two meanings, namely a descriptive and a performative one:

> A person makes a *descriptive* function ascription if she holds or expresses a belief (or similar propositional attitude) that an object has a certain function. Hence, when I tell a friend that a particular object in my violin case is a shoulder rest, I make a descriptive function ascription. A *performative* function ascription is an utterance or other action by which a

[15] The quote from Cummins in section 2.5.1 illustrates this point.

person assigns or tries to assign a function to an object that the object did not have before. A decision to start using a particular cushion as a shoulder rest constitutes a performative function ascription in this sense.

In the literature on functions this distinction is seldom taken into account. However, descriptive and performative function ascriptions are not to be confused. More or less similar to Hansson's distinction I will interpret descriptive function (kind) ascriptions as an epistemic activity: a descriptive function (kind) ascription is making an epistemic claim, the success of which primarily depends on the available evidence. A performative function ascription I will take to be closely related to what Searle calls a function assignment. According to Searle a function F is assigned to an object whenever people (try to) use it as, make it as, treat it as, or think of it as an object with the function F (see the quote from Searle in section 2.5.2).[16] Roughly, a function assignment to an object is successful, if the object is able to perform that function. A more difficult problem is to explicate the role of function assignments in being an instance of a particular technical kind. I will postpone a discussion of this topic until later when we discuss Searle's theory in detail.

To underline the difference between descriptive function (kind) ascriptions and performative function ascriptions, note that the latter may play an important role in epistemic theories of function (kind) ascriptions. For instance, a descriptive function ascription may be based on performative function ascriptions (function assignments): a person A may make a descriptive function ascription to an object X on the basis of her knowledge of a performative function ascription to X by another person B or a some social group. This is exactly what is at issue in mind-dependent theories of function. I will come back to this point in more detail shortly. In order to avoid confusion I will in the following refer to performative function ascriptions as function *assignments*; for the descriptive case I will use the expression function (kind) *ascriptions* or *attributions*.

Epistemic theories of function (kind) ascriptions are intended to explicate function (kind) ascriptions in the descriptive sense. An agent A who is justified in ascribing the property of being for φ-ing (being a φ-er) to X, knows or is justified in believing that X has the property of being for φ-ing (being a φ-er) and vice versa. Note that in general it is possible that A may be justified in *ascribing* a functional property to X independent of A or anybody else *assigning* that functional property to X; in that case we will be dealing with a mind-independent theory of function ascriptions.

The general form of epistemic function (kind) ascription theories presented above is rather simplistic in that it is assumed that function (kind) ascriptions are independent of anything else. As an example of a relational theory of function

[16] Note that Hansson relates a performative function ascription to the assignment of a function to an object that that "object did not have before". By taking over Searle's notion of function assignment, I drop the latter requirement.

ascription consider the ICE-theory as proposed in its original form by Houkes and Vermaas (2003)[17]:

> An agent *a* ascribes the capacity to ϕ as a function to an artifact *x*, relative to a use plan *p* for *x* and relative to an account *A*, iff:
>
> I. the agent *a* has the capacity belief that *x* has the capacity to ϕ, when manipulated in the execution of *p*, and the agent *a* has the contribution belief that if this execution of *p* leads successfully to its goals, this success is due, in part, to *x*'s capacity to ϕ;
> C. the agent *a* can justify these two beliefs on the basis of *A*; and
> E. the agents *d* who developed *p* have intentionally selected *x* for the capacity to ϕ and have intentionally communicated *p* to other agents *u*.

In this theory, the function ascription by an agent to an object *X* is defined relative to a use-plan *p* for *X* and relative to an account *A*. This means that for an object *X* to have the function to φ is a *relational* property. Note that in this form the ICE-theory is not a full blood epistemic theory because of condition E; it does not state an epistemic condition about the beliefs of agent *a*.[18] In later versions this condition is reformulated in an epistemic way, which turns the ICE-theory into a genuine epistemic theory (see (Houkes and Vermaas 2010, Table 4.2, p. 100) and Vermaas (2009)).

If we take into account the possibility that theories of function (kind) ascriptions may be relative to some set of items denoted by *R*, we arrive at the following:

Theory of Relational Function Ascription:

> Agent *A* is justified in ascribing the functional property of being for φ-ing to object *X* relative to *R* iff agent *A* knows that *C1, C2,…,Cm*.

Theory of Relational Function Kind Ascription:

> Agent *A* is justified in ascribing the functional property of being a φ-er to object *X* relative to *R'* iff agent *A* knows that *K1, K2,…, Kn*.

Here it is assumed that the items in *R* and *R'* are the object of some of the beliefs *C1…Cm* and *K1…Kn* respectively. If not, there would be no point in relativizing the ascription of function (kind) to *R* respectively *R'*. *R* (*R'*) may contain various kinds of items, as is illustrated by the ICE-theory, where it contains a use-plan and an

[17] In their 2003 paper Houkes and Vermaas do not explicitly state whether the function ascription of their ICE-theory is to be taken in the descriptive or performative sense. I take their original ICE-function theory to be a descriptive function ascription theory. In later work Houkes and Vermaas (2010) explicitly present the ICE-theory as a descriptive theory.

[18] Condition C may be taken to express a belief of agent *a* by reformulating it in the following way: the agent *a* believes that his capacity belief and contribution belief of condition I may be justified on the basis of *A*.

account. It may also contain items such as social groups (users, designers), social practices or a system of which X is a part.

Ontological theories of function, to which we turn now, are intended to explicate what it means for an object to have functional properties in the ontological sense of 'have'. We assume that functional properties are not among the most basic ontological properties of the world that cannot be further ontologically explicated. So functional properties may be related to or construed in terms of other, more basic ontological features. Most ontological theories of functions explicitly or implicitly make assumptions about a (more) basic ontology of the world and then analyse the ontological status of functional properties against the background of this (more) basic ontology. Taking into account that functional properties may be construed as ontologically relational properties we propose the following general form for ontological theories:

Ontological theory of function:

Object X has the functional property of being for φ-ing relative to S iff X satisfies the conditions $O1,...,Oj$.

Ontological theory of function kind:

Object X has the functional property of being a φ-er relative to S' iff X satisfies the conditions $P1,...,Pk$.

If we assume, as before, that X is a φ-er implies that X is for φ-ing, then the set of conditions $O1,...,Oj$ is a (proper) subset of the set of conditions $P1,...,Pk$. The conditions $O1,...,Oj$ ($P1,...,Pk$) are to be stated in terms of the basic ontological properties of X and some of them have to refer to S (S').

As an illustration of an ontological theory of function that comes close to interpreting a function as a physical property, consider the following, Cummins-style theory[19]:

Object X has the functional property for φ-ing relative to a system S with capacity ψ (i.e. has the function to φ relative to system S with capacity ψ) iff

i. X is part of system S, and
ii. X has the capacity to φ, and
iii. X's capacity to φ contributes causally to S's capacity to ψ.

[19] I call it a "Cummins-style theory" because the analytical account A is suppressed; see Cummins (1975).

In this ontological definition it is assumed that according to the basic ontological structure of the world objects may have capacities. Moreover, the relation "being part of" in (i) and the causal relation in (iii) are taken to be ontological relations.

This Cummins-style theory strongly assimilates functions into the ontology of the physical world. In contrast to this approach, consider McLaughlin's ontological theory of functions. McLaughlin sets out to present an ontological analysis of what it means to be a technical artefact and how an artefact acquires its function (2001, p. 43). He claims that artefact functions are ontologically conferred, attributed or ascribed to objects by agents (he uses these terms indiscriminately).[20] According to McLaughlin, the function of an object is conferred onto the object through the beliefs and desires of an agent. When there are no agents, there are no purposes and therefore no functions. This means that without agents there are no artifactual functions or artifactual categories. McLaughlin (2001, p. 44) claims that "Screwdrivers, tractors, pruning knives are culturally determined functional kinds, not natural kinds." He does not mean his view to imply that functions and function kinds have no place in the ontology of the world. Instead his claim is that in so far functions and function kinds exist they exist relative to the mental states of human agents. Now suppose that these mental states have themselves ontological significance and are part of the basic ontology of the world. Then the following McLaughlin-style ontological interpretation of functions may be proposed:

> Object X has the functional property for φ-ing (being a φ-er) relative to the mental states of agent a iff
>
> i. agent a has mental states which confer on (attribute, or ascribe to) X the functional property for φ-ing (being a φ-er).

Note that in this ontological theory of function (function kinds) the physical capacities of X play no role at all. The reason is that according to McLaughlin criteria for successful use in principle play no role in conferring functions upon objects (see section 2.5.2).

Having distinguished these two kinds of theories of functional properties, an obvious question is how they are related. Leaving aside fundamental issues about how epistemology and ontology in general are (to be) related to each other, I will restrict myself to a few remarks that concern this specific case of function theories. With regard to ontological theories of function it seems important to take into account some form of 'epistemic access' to the ontologically defined functions. What point could there be in introducing an ontological definition of functions such that it would in principle be impossible to have knowledge of these functions?

[20] Note that McLaughlin uses the notion of function ascription in an ontological sense (as opposed to the epistemological sense defined above).

Assuming that we may have knowledge of part-whole relations, physical capacities, and causal relations, the Cummins-style theory satisfies this demand. This demand of epistemic access does not imply that in each and every case where some object *X* ontologically has a function it will be possible to gain knowledge of that function. Suppose that the ontological definition of functions refers to events in the history of *X*, for instance to the intentions of the designer of *X*. Even if in general we may have knowledge of the intentions of other people, situations may occur in which all information about the relevant historic events is lost forever.[21] Then, it may occur that object *X* has ontologically a function, knowledge of which has become impossible. In principle, however, it would have been possible to have knowledge of this ontological state of affairs if knowledge about the historic events involved would have been available. So, depending on general assumptions about what kind of knowledge human agents may have, I take ontological theories of functions to be such that they allow in principle knowledge of those functions. One way to assure this is to construe ontological theories of functions on the basis of the 'ontological commitments' of the most viable epistemic theories of functions.[22]

Given the above elaborations of epistemic and ontological theories, we are now in a position to further clarify our conclusion of the previous chapter that what is needed is a hybrid theory of technical functions. For epistemic theories of function (function kind) this implies that justified function (function kind) ascriptions are to be based on knowledge of or justified beliefs about human intentions and physical features. For ontological theories it means technical functions have to be grounded in human intentions and physical features as basic constituents of the ontology of the world.

3.4 Function assignments and mind-dependent theories of function

Function assignments (performative function ascriptions) play an important role in mind-dependent theories of functions. According to mind-dependent theories of functions, objects have, whether 'have' is interpreted ontologically or epistemologically as 'are justifiably ascribed to', functional properties in relation to human intentions or intentional activities. One form these human intentions or intentional activities may take is function assignments. Searle, for instance, speaks of the assignment of functions and McLaughlin about conferring, attributing or ascribing functions to objects, activities that are all to be taken in a performative sense. According to such mind-dependent theories of functions, objects have functional properties in relation to function assignments by humans. Thus, function assignments

[21] See also Dipert's discussion (1993, p. 15) of what it means for an object to be artifactual.

[22] I put ontological commitments between quotation marks because this notion was originally developed by Quine for formalized theories, whereas here it is used in the context of informal theories; see Quine (1980).

may play a role in epistemic and ontological theories of functions and therefore have epistemological and ontological significance.

Function ascriptions (descriptive function ascriptions), on the contrary, cannot play a role in the conditions *C1, C2,...,Cm, O1,...,Oj* et cetera that figure in epistemic and ontological theories of functions. Reference to function ascriptions in epistemic theories of function would make those theories circular, for those theories are intended to explicate what it means to ascribe (in the descriptive sense) functional properties to objects, that is, to explicate function ascriptions. Any reference to function ascriptions in ontological theories of functions would certainly be problematic for that would imply that epistemic claims about functional properties of objects would play a significant role in the ontological status of those functional properties. This would run against the idea that within an epistemic context the direction of fit is from our beliefs to the world, and not the other way around. Function assignments may figure in ontological theories of function because just as performative intentional acts may have ontological implications for the social world (e.g., the signing of documents may create a new social entity such as a firm), performative intentional acts may play a role in creating new technical artefacts. This is something that technical artefacts have in common with social objects, which underscores our basic claim that technical artefacts have a hybrid, dual nature.

Note that in mind-dependent theories of functions the agent that assigns a function is not necessarily identical to the agent that ascribes a function (i.e. the agent that makes an epistemic claim about an object having a function). This opens up the possibility of what Thomasson calls a realist epistemology with regard to artefacts kinds, even in case functions and function kinds are taken to be mind-dependent (2003, p. 583–584):

> The possibility of members of a group G making substantive discoveries about a certain kind presupposes that it exist and have its nature independently of G member's beliefs and concepts regarding its nature. That, however, does not require that it exist and have its nature independently of *everyone's* beliefs and concepts.

Consider, for instance, an archaeologist who is trying to figure out the function of an artefact of some tribe that, according to an ontological theory of functions, she takes to be determined by the intentions of the makers and users of the artefact. Here the agent who ascribes the function is different from the agent who assigns the function. The archaeologist in the role of 'observer' is engaged in an epistemic activity; she is interested in making (reliable) knowledge claims about the function of the artefact. The makers and users of the artefact may have had primarily only pragmatic interests in the artefact, in the context of which they made function assignments, and no epistemic interest.[23]

---

[23] See also Vermaas and Houkes (2006, p. 8) who distinguish between different perspectives that agents may take with regard to a technical artefact, namely of a user, a designer and a observer. They do not, however, relate their distinction of different perspectives to the distinction between function ascriptions in the descriptive and the performative sense. I take it that the observer perspective is related to function ascription whereas the designer and user perspective are related to function assignment.

One crucial question that remains to be solved concerns which kinds of function assignments may have ontological and epistemic significance. Not just any function assignment will do. I may start trying to use my telephone as a stapler, that is, assign it the function of a stapler, but this is a rather problematic function assignment in the sense that it does not support epistemic or ontological claims that my telephone is a stapler. Constraints will have to be imposed on function assignments in order to warrant their epistemic and ontological significance. What is the nature of these constraints? For the moment I will have little to say on this topic, except that these constraints may be of an epistemic nature, for instance that the assigning agent knows or justifiably believes that the object to which the function is assigned has a particular physical capacity that realizes the assigned function or that she knows or justifiably believes that if used in an appropriate way that object will successfully realize the assigned function. In this way, epistemic considerations concerning the beliefs of the assigning agent may enter into epistemic and ontological theories of function. This does not lead to circularity in the case of epistemic theories of functions, since the beliefs of the assigning agent are different from the believes of the ascribing agent.

It is important to point out, furthermore, that the epistemic constraints involved in function assignments are generally speaking different from the epistemic constraints in function ascriptions and are not to be confused with each other. This may be illustrated with the help of the McLaughlin-style theory of functions discussed above. According to McLaughlin (2001, p. 60) "The truth conditions for artifact function ascriptions involve the beliefs and desires of agents, but they presuppose neither the truth of the beliefs not the rationality of the desires." Indeed, any (descriptive) function ascription will have to satisfy truth conditions which concern justified references by the ascribing agent to beliefs and desires of some assigning agent, since there are no functions without function assignments. From on ontological point of view, however, it makes no difference at all whether the beliefs of the assigning agent are true or her desires rational. This is what makes McLaughlin's ontological theory of functions so permissive. Let us assume that there are no epistemic constraints to be imposed on the beliefs of the agent that assigns a function to an object. Whatever epistemic theory of function ascriptions one would like to add to complement this McLaughlin-style ontological theory, a similar assumption with regard to the epistemic beliefs of agents ascribing a function to that object would lead to the rather absurd result that any function ascription to that object would be as good as any other. More in particular, this means that function ascriptions would in no way be constrained by function assignments, whereas the latter, successful or not, would play a crucial role from an ontological point of view. In fact, the assumption that there are no constraints to be imposed on the beliefs of the ascribing agent, would lead to the conclusion that the development of epistemic theories of functions is a pointless undertaking.

Things become still more intricate when the same agent does the ascription and assignment of a function. That situation occurs when someone creates a first instance of a new kind of technical artefact, for instance, a new kind of corkscrew, and claims that the object she has made is a corkscrew. Now we are dealing with a function assignment and a function ascription by one and the same person. This situation may be interpreted as one in which the creator has a privileged epistemic status in the sense that her claim that the new object is a corkscrew cannot be false, given that her function assignment satisfies certain conditions.[24] Thomasson's theory of artefact kinds, to be discussed in more detail in the next chapter, allows the creator of a new artefact kind such a privileged epistemic status. This special status is related to the fact that the agent involved has direct knowledge of her function assignment. However, depending on the ontological and epistemic theories of function (kinds) adopted, this situation may be interpreted in other ways. The ontological theory of functions, for instance, may include as a condition the *social* assignment of function (which would exclude the possibility of Robinson Crusoe creating a new kind of technical artefact on his island). This ontological view on function (kinds) may be reflected in epistemic theories of function (kinds) such that an agent cannot make a justified function ascription simply on the basis of her own function assignments.

3.5 Theories of technical functions

For a long time discussions about theories of functions have focussed on biological functions, and in so far technical functions were addressed at all it was often only for comparison purposes. Only recently technical functions have started to attract more attention and have been studied for their own sake.[25] In this section I will examine three of the most important theories of technical functions discussed in the literature, namely the ones developed by (1) Searle, (2) Preston and (3) Houkes and Vermaas. Searle's theory of technical functions is more or less a by-product of his overall attempt to show how social reality is related to physical reality. His theory is of particular interest for our project because he discusses in depth how physical, technical and social objects are related. Preston's analysis of technical functions is of interest because she presents an account of the distinction between proper and accidental functions, a distinction that we have come across already several times and that appears to be of crucial importance for theories of technical functions. Finally, Houkes and Vermaas's ICE-theory is a theory of functions that has been developed specifically to account for technical functions, more in particular for the dual nature of technical artefacts; that makes it directly relevant for our own enterprise of clarifying this dual nature.

[24] See, for instance, Thomasson's discussion of our epistemic relation to artifactual kinds in (Thomasson 2007).

[25] For a survey of theories of artefact functions, see Preston (2009b).

3.5.1 *Searle: screwdriver philosophy*

3.5.1.1 The assignment of functions

Searle develops his theory about functions in the context of his attempt "to assimilate social reality to our basic ontology of physics, chemistry, and biology" (p. 41).[26] According to this basic ontology, the world is made up of elementary particles and forces. These particles combine to form systems such as atoms and molecules and some of these are conscious living systems. Consciousness, finally, brings intentionality into the world, that is "the capacity of the organism to represent objects and states of affairs in the world to itself" (p. 7). Objects that belong to the basic ontology of the world have no functions. Technical and social objects do. So, how do they fit into this basic ontology?

To understand how Searle solves this problem we have to turn to his distinction between intrinsic and observer-relative features of the world. Intrinsic features are features that exist independently of conscious observers and their representations of the world, and thus independently of intentionality. For instance, the fact that an electron has a certain mass and a certain charge are intrinsic features of the world. Observer-relative features, on the other hand, exist only relative to the intentionality of conscious observers. The example of the screwdriver (see chapter 2) illustrates that being a screwdriver, i.e., being an object with a certain function, is an observer relative-feature of the world. According to Searle this is true for all functions (p. 14):

> The important thing to see at this point is that functions are never intrinsic to the physics of any phenomenon but are assigned from outside by conscious observers and users. *Functions, in short, are never intrinsic but are always observer relative.*

Humans have, as a matter of fact, the remarkable capacity to assign functions to objects; the only functions there are, are the functions assigned by human beings. So there are no teleological phenomena in the mind-independent world.

It is now possible to see how the ontology of the social world can be grounded in the basic ontology of the world. Since intentionality itself is an intrinsic feature of the world, i.e., is part of the basic ontology of the world, and since intentionality brings with it the capacity to assign functions, functional features of the world are anchored in a specific feature of the basic ontology of the world.[27] Intentionality becomes the linchpin between the physical world and the social world, and enables Searle to construct social reality out of the basic ontology of the world.

Searle does not analyse the notion of function (function assignment) in terms of necessary and sufficient conditions and, as we will see shortly, it is difficult to reconstruct such conditions from his analysis. He mentions two central features related to functions (p. 19). The first is that whenever there is talk of functions,

[26] Page numbers refer to Searle (1995).

[27] That is the reason why Perlman (2004) classifies Searle's theory of functions as naturalistic and non-reducible.

reference is made implicitly or explicitly to a system of values (or purposes or some form of teleology) that humans, as conscious beings, hold. The second is that a normative element is involved in function assignment; when the function of *X* is to φ, then *X* is *supposed to* cause φ. It is not simply causation that is involved in function assignment since an object that malfunctions, that is, does not cause φ, may still be assigned the function to φ. Thus the assignment of functions always involves more than just intrinsic features (causal features) of the world: it brings into play values (purposes, goals) and normative elements.

Searle distinguishes between three different types of function assignment, resulting in three classes of functions: agentive, nonagentive and status functions.

- *Agentive functions.* Searle speaks of agentive functions whenever the functions ascribed to objects refer to the "use to which we intentionally put these objects" (p. 20). Examples of agentive functions are technical functions like screwdriver, bathtub etc.
- *Nonagentive functions.* Nonagentive functions, for instance the biological function of the heart to pump blood, are not assigned because they serve practical purposes, but are assigned to naturally occurring objects in the context of a theoretical account of that object.
- *Status functions.* Finally, there is a special kind of function, which is characterized by the fact that the function of the object is to represent, symbolize or stand for something else; these are status functions and examples of this type are the function of the sentence "Snow is white", but also the function of a landmark or a ten dollar bill.

In the following, I will concentrate mainly on Searle's analysis of agentive functions, since technical functions belong to this type.

The functions of technical objects, like a screwdriver, and of social objects, like a ten dollar bill, fall within the category of agentive functions; their functions are assigned by users to serve practical purposes. But according to Searle, there is an important difference between the two kinds of functions. For objects with technical functions there is a strong link between function and physical structure: they are able to perform their function because they have the appropriate physical structure. These functions are therefore called *causal agentive functions*. The situation with regard to social functions is different. An object with a social function, for instance a ten dollar bill, cannot perform its function on the basis of its physical characteristics. A certain piece of paper can perform its function as money only because a *status function* is imposed on it, and this status function is collectively recognized. So agentive functions come in two different stripes, causal agentive functions and status functions.

A closer look at how status functions are assigned shows the important role of collective intentionality. A status function *Y* may be assigned to a physical object *X* through a constitutive rule, which has the following form: "*X* counts as *Y* in *C*", where *C* refers to a certain context. Thus, status functions as well as technical (causal) functions are assigned to objects but in very different ways: the former are assigned implicitly or explicitly through constitutive rules and can be performed only by virtue of collective intentionality, whereas the latter are assigned on the basis of the intrinsic physical properties of the objects involved. This is the reason why statements about technical functions and status functions have a different character. For Searle, the statements

(i) "This object is a ten dollar bill", and
(ii) "This object is a screwdriver"

have in common that both refer to objectively ascertainable social facts about observer relative features of the world. But the statement about the ten dollar bill is an institutional fact, whereas the statement about the screwdriver is not. The truth of (i) depends on the collective acceptance of a constitutive rule, whereas the truth of (ii) does not.

In spite of the difference in the way causal and status functions are performed, they have an important property in common. According to Searle, for any observer-relative feature *F*, including "being a screwdriver", it is true that seeming to be *F* is from a logical point of view prior to being *F*, that is, seeming to be *F* is a necessary condition for being *F* (p. 13). The meaning of expressions like '*X* functions *as* an φ-er', '*X* is used *as* a φ-er' or '*X* is thought of *as* a φ-er' all contain an element of "*X* seems to be a φ-er" for observers or users of *X*. Whether or not to function as, to be used as or to be thought of as a φ-er is, generally speaking, also a *sufficient* condition for being a φ-er, remains to be seen. But for observer-relative features of the world, such as being a φ-er, some way of seeming to be a φ-er (being used as etc.) is at least a *necessary* condition for being a φ-er.

Let me conclude this overview of Searle's interpretation of technical functions with the final remark that from an epistemic point of view observer relative features of the world, for instance being a screwdriver, may be as objective as intrinsic features, for instance the feature that a hydrogen atom has one electron. Being a screwdriver is ontologically an observer-relative feature of the world and is therefore ontologically subjective. Nevertheless it is according to Searle from an epistemic point of view an objective judgment whether a certain object is a screwdriver or not. It is an objective judgment because "the facts in the world that make [it] true or false are independent of anybody's attitudes or feelings about them" (p. 8). I take this to mean that the ascription (not assignment) of being a screwdriver (being a φ-er) to a particular object *X* by an agent is independent of the ascribing agent's attitudes, feelings or point of view about *X* and that therefore this ascription may be justified on epistemically objective grounds. If this would not be the case, it would be hard to understand how being a screwdriver could be an epistemically objective fact about *X*. Nevertheless, from an ontological perspective this is an observer-relative feature of *X*.[28]

[28] Searle's position with regard to epistemologically objective, but ontologically subjective facts is strongly similar to Thomasson's position on the possibility of a realist epistemology for mind-dependent artefact kinds (see section 3.4).

3.5.1.2 To be or not to be a screwdriver

Searle does not work out an epistemic or ontological theory of function (kind) in terms of necessary and sufficient conditions and leaves many details about the assignment of technical functions to be filled in, which is not surprising since he develops his theory of technical functions as a by-product of his theory of social reality. Moreover he does not distinguish carefully between the assignment of functions in a performative sense, and in the descriptive sense. When he claims that it is an objective feature of an object *X* that it is a screwdriver, he is clearly making a descriptive function-kind ascription claim. However, when he claims that all functions are assigned, assignment is to be taken in the performative sense. Searle also bypasses any discussion about the distinction between proper and accidental function. All of this makes it difficult to get a good grip on his analysis. In the following I will make an attempt to interpret Searle's theory of functions in terms of our general characterizations of epistemic and ontological theories of function and the distinction between function ascriptions and function assignments. I will argue that his theory in its present state does not present a clear view on the role of intentionality in the assignment of functional kinds and that it is not able to come up with a viable distinction between the ascription or assignment of proper and accidental functions. I consider both points to be serious shortcomings of his theory.

Searle mainly deals with the ontological question of when an object *X* is a screwdriver. He is interested in what I have called an ontological theory of function kind. What are necessary and sufficient conditions for being a screwdriver? In order to avoid cumbersome language, expressions like "*X* is a screwdriver", "*X* is for driving screws" et cetera will be taken from now on in their ontological sense; the epistemic counterparts of these expressions will be formulated in terms of the ascription of functional properties. From Searle's analysis it follows that it is a necessary condition for something to be a screwdriver that it is assigned the function of a screwdriver. The statement "Object *X* is a screwdriver" logically implies an indefinite inclusive disjunction of the form "*X* is used as a screwdriver, or *X* is thought to be a screwdriver or *X* is designed as a screwdriver or etc." (p. 32). I will call this the necessary assignment condition and formulate it in the following way:

Necessary assignment condition: X is a φ-er $\rightarrow X$ is assigned the functional property of being a φ-er.

What about a sufficient condition for something to be a screwdriver? What other necessary conditions have to be added to the necessary assignment condition so that together they become sufficient and the implication can be turned around?

On this point Searle has not much to offer. He remarks that in case the agentive function of an object is performed fully in virtue of its physical properties (p. 45):

> ...we do not have any metaphysical doubts about whether or not this is really a screwdriver, or this is really a car, because the sheer physical features of the objects in question enable them to function as screwdrivers or cars.

It is not so clear how this passage is to be interpreted. It suggests that if an object X has the appropriate physical capacities to drive screws, then this is sufficient for the epistemic ascription of being a screwdriver (without any metaphysical doubt). The ontological counterpart of this would lead to the following sufficient capacity condition:

Sufficient capacity condition: If X has the physical capacity to $\varphi \rightarrow X$ is a φ-er.

For at least two reasons such a sufficient capacity condition would be problematic.

In the first place, it leads to a conception of function kinds that is much too liberal. Objects, such as a coin, may have the appropriate physical properties which allow them to be used as and thus to function as screwdrivers, without really being screwdrivers. Apparently Searle is aware of this, since in a footnote (p. 53–54) he remarks: "You could not define "screwdriver" as "anything that can be used as a screwdriver," because lots of things can be used as screwdrivers that definitely are not screwdrivers, for instance, coins." This remark is in line with his point of view that seeming to be a φ-er (being used as a φ-er, being designed as a φ-er, being thought of as a φ-er) is only a necessary condition for being a φ-er.

Searle's remark about the coin makes clear that he assumes that not every successful assignment of a function to φ to an object X implies that X is a φ-er. A coin may be successfully assigned the function of a screwdriver, but that does not make it a screwdriver. In order to solve the counter-example of the coin, he will have to distinguish between *the assignment of functions that imply that the object involved is an instance of the corresponding functional (technical) kind, and assignment of functions for which this is not the case*. This is a problem he does not address. It is an issue that appears to involve the distinction between the assignment of proper and accidental functions; at first sight only the assignment of a proper function implies that the object involved is an instance of the corresponding functional kind. Scheele (2006) has proposed that Searle may deal with this problem by treating proper functions as status functions. That brings me to the second reason why the sufficient capacity condition is problematic. It concerns the role of intentionality in the assignment of causal functions.

The idea that having certain physical properties is sufficient for something to be a screwdriver runs counter to Searle's idea that the *assignment* of the function of driving screws is a necessary condition for being a screwdriver. It makes being a screwdriver an intrinsic, observer-independent feature of the world, whereas Searle maintains that it is an ontologically subjective, observer-relative feature. What precisely is the role of the assignment of functions, and therefore of (collective) intentionality, with regard of causal functions? In what sense can it be claimed that causal functions are assigned to objects, given that the successful performance of these functions does not so much depend on (collective) intentionality as well on the physical properties of the carriers of these functions? Put in another way, is there anything inherently intentional about the feature that X is a screwdriver, since the crucial point appears to be that it can perform the function of a screwdriver on the basis of its physical makeup?

According to Searle, the assignment of causal functions may be a matter of individual or collective intentionality (p. 38–39; 122), but either way leads him into trouble. Let us start with collective intentionality. In the course of his analysis of the nature of constitutive rules he makes a remark that downplays the role of *collective* intentionality for the assignment of causal functions. Constitutive rules of the form "X counts as Y in C" allow attaching a status function to an object through collective intentionality. But not every rule of the form "X counts as Y in C" is a constitutive rule (p. 44):

> Furthermore, it does not express a constitutive rule to say "objects of a certain shape count as chairs," because the functions assigned can be assigned independently of any human agreement. If it has a certain kind of shape, we can use it as a chair regardless of what anyone else thinks.

Causal functions, thus, are not assigned through constitutive rules, because they can be assigned "independently of any human agreement" whereas constitutive rules require collective acceptance. So, *collective* intentionality is not necessary for the assignment of causal functions.[29] This means that the fact that *X* is a screwdriver is not necessarily a social fact. By stipulation, a social fact involves collective intentionality; all and only cases involving collective intentionality are social facts (p. 122). Nevertheless it is classified as a social fact in Searle's hierarchical taxonomy of kinds of facts. Searle's footnote about a coin not being a screwdriver suggests that function assignment cannot be a matter of individual intentionality either. On the basis of its sheer physical properties the coin may be assigned the function of a screwdriver by an individual. That assignment, however, does not make it a screwdriver. The most obvious reason is that we are dealing here with the assignment of an accidental function and not a proper function. Again we run up against the problem of distinguishing between assignments of accidental functions and of proper functions. In order to solve this problem an appeal to the social context, that is, to some element of human agreement or collective intentionality, appears to be unavoidable. That would suggest that proper causal functions are status functions that are related to constitutive rules. The above quotation from Searle, however, seems to undermine the possibility to interpret proper functions as status functions.

In conclusion, Searle's analysis of causal agentive and status functions explicates a fundamental difference in the way technical and social artefacts perform their function. However, it is highly doubtful whether his analysis can be supplemented with a coherent ontological theory of technical functions and technical kinds (and a corresponding epistemic theory) in terms of human intentions and physical features. The necessary assignment condition focuses on the ontological role of human intentionality and the sufficient capacity condition on the ontological role of physical features. Even if these two conditions would be compatible within Searle's theory, which appears doubtful, it is not a straightforward matter how these two conditions could be combined into a viable theory of technical function (kind). Adding to the

[29] See also Thomasson (2003, p. 599, footnote 25).

necessary assignment condition the necessary condition that the object involved must have the appropriate physical capacities to φ leads into a dead end. This does not work because of malfunctioning technical artefacts. A malfunctioning TV-set is still a TV-set, but it does not have the appropriate physical properties for performing its function. Thus, having the appropriate physical conditions for φ-ing is certainly not a necessary condition for being a φ-er. Moreover, together these necessary conditions are not sufficient for an object to be an instance of a technical kind as the example of the coin illustrates. As it stands, Searle's theory does not offer a solution to the problem of how to combine intentional and physical features into a coherent theory of technical functions and technical kinds.

3.5.2 Preston: proper functions

Preston (1998; 2000) discusses two distinct notions of functions, each of which she considers to be relevant for analysing technical (as well as biological) functions, namely systems functions and proper functions. The main difference between these two notions is that system functions are tied to actual capacities or dispositions of objects whereas proper functions are related to their selection history. System functions are the kind of functions considered by Cummins (1975). An object has a system function in relation to the system in which it is actually embedded; historical considerations play no part in analysing system functions. Since an object may be embedded in different systems during its lifetime, it may acquire different systems functions in the course of its existence. System functions may therefore be very volatile; they are "constantly coming into and going out of existence" (Preston 1998, p. 250). The proper function of an object, according to Preston, is determined by the selection history of that object. Its proper function is that capacity or disposition for which its ancestors have been reproduced in the past. For technical artefacts, ancestors are defined as objects of the same sort, kind or type as the object to which the proper function is attributed. Artefact kinds may be said to have proper functions in so far their members have proper functions. Proper functions of artefact kinds cannot be lost or changed easily because they belong to the kinds in virtue of their selection history. They are related to lineages of things (see also (Neander 1991, p. 174) and can only be changed by changing these lineages (an artefact kind may change its proper function when objects of this kind are reproduced, without any significant change, because they perform a new function (Preston 1998, p. 248; Preston 2000, p. 31)).[30]

[30] On this point there is some tension in Preston's analysis. She remarks that (1998, p. 247) "This replacement of one proper function with another is common among artifacts", which suggests that it may be easy to change proper functions of artefacts. However, in discussing 'ongoing system functions' (that is, system functions that do not disappear but also are not transformed into proper functions by selection) she remarks that changes of ongoing systems functions into proper functions are rare (1998, p. 241). But since all new proper functions start out as (ongoing) system functions, this seems to imply that changes of proper functions are rare, since they involve (1998, p. 248) a "wholesale reproduction only for that new function."

Whereas systems functions are not normative, proper functions are: if an object with a proper function malfunctions, then it makes sense to say that it is supposed to or ought to perform its function. This is one of the main differences between the two kinds of functions. According to Preston, these distinct notions of functions are not rival notions of functions but equally important, complementary elements for a viable general theory of technical functions. Proper functions stress the *stability* of functions over time and groups of things, system functions their *lability*. The distinction between proper and system functions is one between what you are supposed to do with a thing and what you can do with that thing. What Preston refers to as system function is more or less the same as what we have referred to so far as accidental function; a flatiron acquires the system or accidental function of a doorstop when it is used to keep open a door.

Preston maintains that technical artefacts have proper functions and that there is a strong analogy between the ways biological items and technical artefacts acquire their proper functions (1998, p. 243):

> Artifacts have proper functions. These are the functions we are most likely to describe in answer to questions like 'What is that for?' or 'What is that?', and they are often reflected in our common names for things – for example, vegetable peeler, light switch, driveway, screwdriver, soap dish, clothes hanger, bookmark, and so on. Artifacts get these proper functions by a process analogous in basic respects to the natural-selection process by which biological traits get theirs.

Successful use of artefacts will lead to their reproduction, just as the positive contribution of a new biological trait to reproductive success will lead to its reproduction. In case of technical artefacts, the history of reproduction typically involves intentional human action that plays a role in the context of their production and distribution, but also in market competition and in their use; socio-cultural selection takes the place of natural selection.[31] She does not consider the fact that certain aspects of this process of reproduction are intentional as opposed to the biological one to make a real difference with regard to the issue of acquiring proper functions. Neither does the fact that in the case of artefacts the reproduction history may be based on *perceived* success as opposed to real success. She discusses the example of electric bug zappers that are supposed to kill mosquito's (1998, p. 245–246). Apparently they kill a lot of insects but only a few mosquito's and on some accounts they even attract more mosquito's than they kill. So, artefacts may be reproduced on the basis of illusory success, that is, for the wrong reasons. In that case we are dealing with what Preston calls 'phantom functions'. As long as it is assumed that successful performance establishes the real proper functions (in the case of the bug zappers: the killing of insects) the analogy with the biological case, she claims, stays valid.

[31] Others have also proposed the idea of socio-cultural selection taking the place of natural selection in the case of technical artefacts; see for instance Millikan (1984) and Bigelow and Pargetter (Bigelow and Pargetter 1987).

For Preston, therefore, technical artefacts acquire proper functions in basically the same way as biological traits.[32] All these proper technical functions, however, start out as system functions; they may become proper functions when these systems functions lead to a suitable reproduction history. This does not mean that artefacts with a proper function are always used in accordance with their proper function. They may perform all kinds of system (accidental) functions. According to Preston (2006) proper functions are co-determined by social use and social reproduction. Together with the assumption that proper functions determine artefact kinds, it follows that artefact kinds are partly socially constituted: for an object to be an instance of an artefact kind becomes a social-historical fact.

In more recent work Preston (2009a) has tried to spell out in more detail the role of use and reproduction with regard to proper functions of technical artefacts. In analogy to attempts to couple proper functions in biology to the notion of fitness, and not to that of selection, she starts off with the following definition of artefact proper functions (Preston 2009a, p. 46):

> A current token of an artifact type has the proper function of producing an effect of a given type just in case producing this effect contributed to the intended use of past tokens of this type of artifact, and thereby contributed to the reproduction of such artifacts.

She argues that this proposal does not work since it cannot account for the difference between manifest and latent functions, the former being grounded in conscious and deliberate use, the latter in unconscious and unintended use. This problem may be remedied by simply dropping 'intended' in the definition. Other problems remain such as how to deal with phantom-functions, that is, cases of proper functions that have perfectly normal histories of use and reproduction, except that the tokens of the artefacts are unable to perform the alleged function (think of the function of the above bug zappers). In contrast to her earlier position on phantom functions, she now opts for a definition of proper functions that allows for phantom functions to be real proper functions. After considering various modifications of the above definition, however, she comes to the conclusion that there is no good analogue of the notion of biological fitness for technical artefacts and that therefore contribution to successful past performance does not single out the proper functions of technical artefacts.

How then to define proper functions of technical artefacts? Preston thinks that only by looking at patterns of actual artefact use and their role in reproducing artefacts will it be possible to identify proper functions. She ends up with the following provisional proposal (2009a, p. 48):

> A current token of an artifact type has the proper function of producing an effect of a given type just in case producing this effect contributes to the explanation of historically attested, dominant patterns of use to which past tokens of this type of artifact have been put, and which thereby contributed to the reproduction of such artifacts.

The details of this proposal still have to be worked out, but Preston is aware of one possible objection, namely that this definition does not allow prototypes to have

[32] A similar position has been defended by Griffiths (1993).

proper functions. She thinks that that is not a problem; according to her prototypes, against common opinion among function theorists, have only system functions.[33]

Preston rejects attempts to define proper functions of prototypes, or technical artefacts in general, in terms of intentions of designers, because she thinks that there are no relevant differences between the intentions of designers and users (Preston 2003, p. 608). She considers four such possible differences: (1) some special cognitive structure of the intentions of designers in comparison to the intentions of users, (2) a difference in creativity between designers and users, (3) the involvement of the intentional modification of things in the design of proper functions whereas there is no such modification in the use for accidental functions, and, finally, (4) the actual content of the intentions of the designer fixes whether a function is a proper or accidental function. She argues against each of these possibilities and concludes that if proper functions are derivable from the intentions of designers, they are also derivable from the intentions of the users. That, however, would obliterate the distinction between proper and accidental functions, because any accidental use of an artefact would lead to a new proper function.

In comparison to Searle's theory of functions Preston's has the advantage of facing the issue of the distinction between proper and accidental (systems) functions head on. She tries to account for the difference in normative impact of attributing proper and accidental functions and explains why accidental functions are much more volatile than proper functions. However, I consider her account of this distinction to be problematic. My main concern with Preston's theory relates to her grounding the proper functions of technical artefacts in their use and reproduction histories, be it the use and reproduction histories themselves, as in her earlier work, or in the explanation of use and reproduction histories, as in her more recent work. In doing so, she appears to inherit a feature of theories of biological functions that, in my opinion, is highly problematic when it comes to theories of technical functions. The way variations on which selection operates come about does not play a significant role in the theory of biological evolution; the variations are random with regard to the selection environment and come into being without proper functional features. It is only later on in their selection histories that variations acquire proper functional features. In line with this, Preston maintains that prototypes of new technical artefacts (new technical variations) do not have proper functions; they acquire proper functions only later on when they are used and reproduced. So any attempt to ground the proper functions of prototypes in the intentions of designers is a senseless undertaking. The use and reproduction approach forces her to deny that the ideas and work that go into the design, development and making of new technical artefacts are relevant when it comes to the proper functions of (prototypes of) technical artefacts.

[33] I interpret Preston's definition as saying that a proper function is attributed to a token of an artefact type *relative* to an explanation of a history of use and reproduction of artefact tokens of that type. This makes proper function attribution a relational affair.

To make clear why in my opinion a use and reproduction approach to proper functions of (prototypes of) technical artefacts is problematic, I will compare two situations, one in which a natural object, a shell, is used for a practical purpose, for drinking water, with one in which a more complex technical artefact such as a video recorder is designed and made. A shell may be used incidentally for drinking water, in which case it acquires temporarily a system function. If that shell turns out to have a hole in it, then it cannot be said to be malfunctioning. Suppose, however, that a social practice develops in which these shells are 'reproduced' (collected or cultivated), traded and sold and used for drinking water. Then these shells acquire on Preston's account the proper function of drinking cups and assuming that proper functions determine technical kinds, these shells are drinking cups. These proper functions are normative: shells (drinking cups) are supposed to or ought to show certain behaviour. So, in case some shell (drinking cup) has a hole in it, it may be said to malfunction. Compare this case to the one of an unsuccessful video recorder, the one developed by the Philips company, called the video 2000. This technically very sophisticated video system, whose picture quality was allegedly better than of its competitors, was a market failure; it was produced and put on the market only for a short time (the main reason apparently being that the video recorder industry adopted a different standard). Do these video recorders have a proper function? On the use and reproduction account this could be claimed to be the case, because of its, albeit short, history of use and reproduction. Suppose, however, that this video-recorder never made it to the market, but that a first series of prototypes was produced that ended up being stored in some basement. Do they have a proper function? On the use and reproduction account they have no proper functions, for there is no history of reproduction. Because they lack the relevant proper functions, these objects stored in the basement are not even video-recorders (again, assuming that proper functions determine the technical kind). This seems rather odd. From an engineering point of view it makes perfectly sense to make normative statements about them: they are supposed or ought to reproduce TV-images; if not, they malfunction and may be repaired (the notion of repair would make no sense if these video recorders would have only an accidental function). Even without a history of use and reproduction, these video recorders appear to have a proper function.

The following considerations further support the claim that technical artefacts may have a proper function without a history of use and reproduction. Take one-of-a-kind technical artefacts or systems like the Oosterschelde-dam, a major accomplishment of civil engineering protecting part of the Netherlands from flooding by the sea, or the Hubble telescope and similar unique scientific instruments, or also relatively simple artefacts like a particular mechanical construction used only once in the construction of some building and scrapped afterwards. Such technical artefacts have a history of use, but not of reproduction. I find it hard to accept that they have no proper functions because they lack a history of use and reproduction (see also (Millikan 1999, p. 205). The Oosterschelde-dam surely has a proper function, and it makes perfectly sense in case it fails to perform its function to claim that it was supposed to protect the hinterland. Suppose the unique mechanical construction fails and some workers get injured or killed; it surely makes sense to refer to its proper function and to make normative claims about what the construction was supposed or ought to do.[34] All in all, on the basis of everyday as well as engineering functional discourse I take it to be a phenomenological fact about functional properties that also one-of-a-kind technical artefacts and prototypes have proper functions. Contrary to Preston I assume more generally that technical artefacts without a historic lineage of use and reproduction may have proper functions.

Let us have a closer look at the different interpretations of the case of the shell and of the video-recorder. In my opinion there is a crucial difference in the kind of activities involved in the design and making of the technical artefacts in the two cases, a difference that Preston denies there is. In the case of the shell, the only activity required is of a mental nature, namely the idea to use a shell for drinking water, to use it as a drinking cup. It may be questioned whether this mental 'design' activity turns the shell into a real artefact, since the material thing itself is left unchanged; its structural or form features remain the same.[35] However that may be, this use requires a creative mental act. This design activity by itself does not bestow a proper function on the shell in case of a creative *accidental* use. A clear sign of this is that this design activity does not by itself warrant normative malfunction statements. Given a suitable history of use and reproduction, it may be granted that these shells acquire a proper function in analogy to exaptations in biology, in which case normative statements with regard to shells do make sense. Whether or not these shells are indeed instances of the technical kind 'drinking cup' is an issue I find difficult to decide; here we are dealing with a limiting case of making technical artefacts and I see no conclusive reasons for arguing either way.

With regard to the video recorder Preston would claim that the situation is more or less similar to the one of the shell. She argues that there is, from the point of view of the issue of proper functions, no essential difference between creative use and creative design. This means that for her the design and making of the video recorder is in principle similar to the creative use of the shell; any difference there is, is only a matter of degree. Consequently, the video recorder does not have a proper function because of a lack of a use and reproduction history. I disagree with the idea that there is no relevant difference between designing and using and with the consequence she draws from that. The story to be told about the making of the video recorder involves not only creative ideas but also, and much more so, various kind of physical activity necessary for creating the appropriate material things (including the tools to produce these things). To describe this kind of design as in no way significantly different from the design of the shell as a drinking cup appears to me untenable. The latter may indeed by characterised as a form of creative using but the

[34] See also Vermaas and Houkes (2003).

[35] According to Hilpinen (1992, p. 69) the adoption of natural objects for practical purposes may be considered a limiting case of making an artefact.

description of the design of the video recorder in that way appears completely off the mark.[36] The making of the video recorder involves apart from creative ideas, the creative manipulation of matter. Making a video recorder is more than only a mental activity, more than only projecting a function to an already existing object, or appropriating an given object for performing a certain function. On the contrary, the situation with regard to the design and making of a video recorder may be characterized as the mirror image of the one with regard to the shell: the desired functional features are more or less given and the aim is to create a material/physical construction that will realize those functional features (see chapter 5). This difference I take to be highly relevant for the issue of proper functions.

To conclude our discussion of Preston's approach to functions, her interpretation of the proper-accidental distinction is of limited applicability since it presupposes a use and reproduction history of the technical artefacts involved. This means that it cannot provide a foundation for the notion of proper function of prototypes and one-of-a-kind technical artefacts; she solves this problem by denying that these technical artefacts have proper functions. Taking into account the mental and physical work that goes into the design and development of technical artefacts in modern engineering practice I assume that these technical artefacts do have proper functions. Preston's interpretation of the notion of proper function appears viable for cases of technical artefacts with a history of use and reproduction practices. It has to be supplemented, however, with a theory that can account for proper functions in the case of the creation of the first instance of a new kind of technical artefact and the creation of one-of-a-kind technical artefacts. In the next chapter I will propose a theory that grounds proper functions in such cases in human intentions (of designers or makers) and in the physical features of technical artefacts. This theory is not intended to replace Preston's theory in cases where technical artefacts have use and reproduction histories, but to be put alongside hers, since, as I will argue there, we are dealing with different notions of proper functions. This brings me, finally, to a point of critique on Preston's approach that will be addressed in the next chapter. It concerns the fact that she, in line with many function theorists, assumes that the proper function of a technical artefact determines its technical kind (see for instance (1998, p. 237). However, this option is not open to her on pain of circularity; this is related to the fact that her definition of proper function is based on the type-token distinction and this distinction itself is based implicitly on the notion of proper function (I will come back to this point in more detail in section 4.2). Precisely by distinguishing different notions of proper functions, this problem may be resolved.

[36] Of course, design engineers may be said to 'use' physical objects and processes when creating the video recorder in the sense that they simply rearrange existing objects. In this fundamental sense, however, also artists are not creating anything but using pre-existing things and all human action on matter would be a form of use.

3.5.3 Houkes and Vermaas: the ICE-theory

The last theory of functions to be reviewed here is the ICE-theory proposed by Houkes and Vermaas (2003, 2006; 2010). This theory is particularly intended to account for technical functions and for the dual nature of technical artefacts. The following outline of this theory is mainly based on its most recent formulation in their book *Technical functions; on the use and design of artefacts* (Houkes and Vermaas 2010).[37] In it they take an action-theoretic approach to analysing the technical functions of useful material, which not only includes engineered technical artefacts but any material things that may be useful for achieving practical goals, up to natural objects. They reconstruct the using and designing of technical artefacts in terms of the central notion of a use-plan, which is "a more or less standardised way of manipulating objects in order to realise a practical goal" (p.8). These use-plans are subject to various standards, such as goal consistency, means-end consistency and belief consistency (p. 37 ff). Technical artefacts are embedded in use-plans and it is only in relation to use-plans that technical artefacts have functions. The use of technical artefacts then amounts to the execution of a use-plan. Designing is also interpreted in terms of use-plans; it is not primarily the creation of technical artefacts or their blueprints, but the creation of use-plans. These use-plans are then communicated to users in order to assist them in realising their goals. The designing of use-plans only secondarily involves the design of technical artefacts in case one of the objects to be manipulated according to the use-plan does not yet exist. In that case designing use-plans also involves what they call 'product designing'. It is against this action-theoretic background that Houkes and Vermaas construct their theory of technical functions.

Applying the method of 'conceptual engineering' they start off, as design engineers would do, with fixing the specifications or list of requirements that the theory of technical functions to be designed must satisfy. To do so they list their intuitions and describe the phenomenological data about technical functions to which they appeal. On the basis of this they come up with the following desiderata that concern four aspects of the phenomenology of artefact use and design, namely use versatility, possible lack of success, physical restriction and innovation (p. 5)[38]:

> *The proper-accidental desideratum:*
> A theory of artefacts should allow that artefacts have a limited number of enduring proper functions as well as more transient accidental functions.
> *The malfunctioning desideratum:*
> A theory of artefacts should introduce a concept of a proper function that allows malfunctioning.

[37] In the following, page numbers without years refer to Houkes and Vermaas (2010).

[38] It is not a theory of artefacts, as stated in the quote, that should satisfy these desiderata but a theory of functions; in the accompanying text this is stated explicitly. Only at the end of their book Houkes and Vermaas address problems concerning a theory of artefacts.

> *The support desideratum:*
> A theory of artefacts should require that there exists a measure of support for ascribing a function to an artefact, even if the artifact is dysfunctional or if it has a function only transiently.
> *The innovation desideratum:*
> A theory of artefacts should be able to ascribe intuitively correct functions to innovative artefacts.

These are the desiderata which Houkes and Vermaas want their theory of technical functions to meet. In one form or another, we have already encountered these desiderata in our discussion of function theories. Of course, the choice of these desiderata and their particular formulations may be questioned, but the only way to disagree productively with their choice is to come up with another list and a competing theory of functions that will convince the 'users' of these theories. That is a challenge I will take up in the next chapter; for now I will refrain from commenting on this list of desiderata and will accept it as it stands.[39]

As a stepping stone to the construction of their own theory of technical functions Houkes and Vermaas evaluate existing theories of technical functions against these desiderata (2003; 2010, Ch. 3). In their survey they leave the proper-accidental desideratum out of consideration since it can be satisfied by their use-plan approach (p. 49), a point to which I shall return shortly. They consider three kinds of archetypical function theories out of which all current function theories may be constructed in one way or the other. First there are the intentional (I) function theories, which characterize functions as intended effects. The functions of technical artefacts are determined by the intentions, beliefs and actions of agents, who may either be designers or users (or other types of agents) (p. 50). According to Houkes and Vermaas I-theories may easily satisfy the malfunctioning and the innovation desiderata, but they flounder on the support desideratum. Because the intentions and beliefs of agents are sufficient for ascribing functions to objects, the physical properties of the objects involved appear not to matter. Next they consider causal-role (C) function theories. These theories interpret the function of an object x in terms of the causal role that that object plays in a larger encompassing system. Suppose that system s has the capacity to ϑ. Then, relative to an analytical account A the capacity to φ of x is its function in s if A adequately accounts for the capacity of the system s to ϑ by appealing in part to the capacity of x to φ in s. According to Houkes and Vermaas, C-theories are able to satisfy the support and innovation desiderata, but unable to account for malfunctioning. Finally, there are the evolutionist (E) function theories, according to which the functions of technical artefacts are determined by their evolutionary history, much like biological functions are. Assuming that an artefact x has a series of predecessor artefacts, the capacity of x to

[39] Note that there are considerable changes in the formulation of the above list of desiderata when compared to the list presented in Vermaas and Houkes (2003, p. 265-266), especially with regard to the support (or physical structure) desideratum. From the point of view of the method of conceptual engineering such changes are to be expected: it is common practice that design specifications are changed during a design process.

φ is its evolutionist function if and only if this capacity contributed positively to the reproduction of its predecessors and of *x* itself (p. 61). One of the main problems of E-theories is meeting the innovation desideratum; it has difficulty in explaining that the first instance of a new kind of technical artefact may have a function, since it lacks predecessors. We have already come across variants of these archetypical function theories in our discussions so far.[40]

On the basis of their assessment of existing theories of functions Houkes and Vermaas propose their own theory, called the ICE-function theory, because it incorporates elements of the intentional (I), causal role (C) and evolutionary (E) theories. It is a theory that "falls squarely in the intentionalist tradition" (p. 3) and builds upon their use-plan analysis. They present the ICE-theory as a function *ascription* theory (p. 78; see also p. 48–49): "It explicitly characterizes justifiable *ascriptions* of functions by agents on the basis of their beliefs and actions, instead of defining functions as *properties* that artefacts have independently of beliefs and actions" (see also Vermaas and Houkes (2006, p. 8)). In order to clarify whose beliefs, intentions and actions matter and who ascribes functions to objects they use their distinction, developed within their use-plan approach, between various roles that agents may play with regard to technical artefacts. They distinguish the roles of designer, justifier, passive user, observer and (technical) analyst. For each of these agent roles they analyse in detail how they ascribe technical functions to objects.

Here, I will focus mainly on the analysis of function ascriptions by designers, and passive users. According to the ICE-theory, function ascriptions by these agents have the following form (p. 100: Table 4.2)[41]:

Function ascriptions by designers or justifiers:
A designer *d* or justifier *j* justifiably ascribes the physicochemical capacity to φ as a function to an artefact *x* relative to a use plan *p* for *x*, and relative to account *A*, iff:

I. *d*/*j* has the belief *Bcap* that *x* has the capacity to φ;
   d/*j* has the belief *Bcon* that *p* leads to its goals due to, in part, *x* 's capacity to φ; and
C. *d*/*j* can justify *Bcap* and Bcon on the basis of *A*.

Function ascriptions by passive users:
A passive user *u* justifiably ascribes the physicochemical capacity to φ as a function to an artefact *x* relative to a use plan *p* for *x*, and relative to testimony *T*, iff:

I. *u* has the belief *Bcap* that *x* has the capacity to φ;
   u has the belief *Bcon* that *p* leads to its goals due to, in part, *x* 's capacity to φ;
   u believes that a designer *d* or justifier *j* of *p* has *Bcap* and *Bcon*;
C. *u* can justify *Bcap* and *Bcon* on the basis of *T*;
   u can justify on the basis of *T* that *d*/*j* has *Bcap* and *Bcon*; and
E. *u* received *T* that *d*/*j* has *Bcap* and *Bcon*.

[40] In their review Houkes and Vermaas (2010, section 3.5) also examine theories that are combinations of these basic theories, but find them all lacking.

[41] Note that whereas the ICE-theory is a theory of function ascriptions, the caption of Table 4.2 reads "The three ICE-definitions for functional descriptions." More often in their book (see for instance p. 88, 99, 121) Houkes and Vermaas appear to make a distinction between functional *descriptions* and function *ascriptions*. I will come back to their notion of function ascription shortly.

The main difference between function ascriptions by designers and justifiers on the one hand and passive users on the other concerns their epistemic resources. Designers and justifiers can rely on an explanatory account of the capacity and contribution beliefs, *Bcap* and *Bcon*, which ground their belief of the effectiveness of the use-plan *p*. Passive users, by contrast, have to rely on testimony *T*. So the beliefs in the effectiveness of a use-plan *p* may be based on various kinds of evidence, ranging from sophisticated scientific and technological explanations to experience of successful use, to reliable testimony by other users or designers.

The conditions in the definitions are labelled I, C and E to show their relation to the three archetypical function theories. These conditions are to be taken as an integral part of the use-plan conditions for being a designer and passive user.[42] According to Houkes and Vermaas the intentional element in their use-plan approach to technical functions resides in the fact that (p. 79) "an artefact is only a means to an end if the agent believes that execution of a use-plan for the artefact realises that end." Only then the artefact may be ascribed a function. But not any function may be ascribed. Here the causal-role element in the use-plan comes into play; the effectiveness belief of a useplan has to be justified. Finally, the evolutionist element in the ICE-theory is to be found in the historical aspect that the use-plans of artefacts have to be communicated to users.

Leaving further details about the ICE-theory aside, I now turn to Houkes and Vermaas' assessment of the ICE theory against their four desiderata. Not surprisingly they claim that the ICE-theory meets all four criteria, although they admit that there is one catch (p. 93–94). First, the proper-accidental desideratum; it is taken care of by distinguishing between proper-use and improper-use plans for technical artefacts. Proper-use plans are plans for artefact use that are accepted within a certain community and improper-use plans correspond to use that is socially disapproved. In combination with the ICE-theory this leads to the characterization of proper-function ascriptions as function ascriptions relative to proper-use plans and of improper-function ascriptions relative to improper-use plans. The innovation desideratum is also met because the historical aspect built into the ICE-theory does not require a series of predecessors but can be limited to the history of the design process. So it is possible to ascribe functions to novel artefacts. Also the support desideratum does not pose any problems. Because of the C-conditions agents must be justified in believing that the artefact has the appropriate capacity that they ascribe to it as its function; so, not any function ascription is justified.

An assessment of the ICE-theory with regard to the malfunctioning desideratum turns out to be not as straightforward. It appears that the theory can account for some forms of malfunctioning, but not all of them. The C-conditions in the ICE-theory state that the ascribing agent must be justified in having the capacity and contribution beliefs. These beliefs need not be true, which leaves open the possibility that, although the agent is justified in having these beliefs, the artefact does in fact not have the relevant capacities. Then we are dealing with a case of a malfunctioning artefact to

[42] For a definition of the roles of designer and passive user, see their Table 4.1 (p. 84).

which an ICE-function can be ascribed. So, suppose that, without me knowing, the on/off switch of my TV-set, which I have been using successfully for a long time, has become broken since my last use. In that situation I can ascribe, on the ICE-theory, the function to produce TV-images to my TV-set, in spite of the fact that it is unable to perform that function. However, as soon as I know that the on/off switch is broken, I can no longer ascribe that function to my TV-set, since I am no longer justified in believing that my TV-set has the appropriate capacity.

Such 'known-to-be-broken artefacts' or 'post-hoc malfunctioning' pose a problem to the ICE-theory, but according to Houkes and Vermaas it is possible to account for such counterexamples without amending the theory itself. In a nutshell, they propose to deal with these cases by introducing a distinction between an artefact *having* a capacity and *exercising* it (p. 106 ff). They challenge the assumption that the belief that an artefact does not exercise a capacity entails the belief that that artefact does not have that capacity. As an example they mention a car that has run out of petrol; it is not able to exercise the capacity to transport people or goods, but still has this capacity. More generally, any broken artefact that is not 'beyond repair' technologically or economically, may be said to have the relevant capacity without being able to exercise that capacity. So they argue that the statement (p. 108) "an agent justifiably believes that an artefact has the capacity to φ and justifiably believes that it does not exercise this capacity" does not contain mutually inconsistent beliefs. This means that in the case of known-to-be-broken artefacts an agent can still satisfy the C-condition (justifiably believe that the artefact has the appropriate capacities) and thus ascribe an ICE-function to the artefacts.

In the final chapter of their book Houkes and Vermaas address some issues concerning the metaphysics of artefacts. They conclude that their analysis shows that technical artefacts have a twofold dual nature (p. 11). One dual nature concerns the fact that technical artefacts have intentional and physical features, which is reflected in the I- and C-conditions of the definition of ICE-functions. This is the kind of dual nature that I have been discussing so far. The other dual nature concerns the fact that technical artefacts are objects used and made. This last duality leads them to the following definition of artefacts (p. 158):

> **A dual definition of artefacts:**
> An object x is an artefact a of type t if and only if: (1) x has been intentionally produced by an agent m; and (2) x is manipulated in the course of executing a specific use plan p, which is designed, communicated and evaluated in accordance with the use-plan analysis of using and designing.

Houkes and Vermaas' theory of use-plans and ICE-functions is a detailed analysis of the second feature of artefacts, that is, artefacts as objects of use. What is still lacking for the metaphysics of artefacts is, they remark, an equally detailed theory of making technical artefacts.

Houkes and Vermaas' analysis of use-plans and ICE-functions is without any doubt the most comprehensive and in-depth analysis of the use of technical artefacts and their functions available. They convincingly argue that functions of technical artefacts and their components are intimately connected to use-plans; without use-plans no ICE-function ascriptions. That is fully in line with the conclusion we reached at the end of Chapter 2 that the functions of technical artefacts are related to a context of human action. One of the merits of their work is that they unpack this context of human action in great detail and among other things show that it hides agents in very different roles with regard to technical artefacts and that this may lead to different ways of ascribing functions to technical artefacts. However, when it comes to explicating the dual nature of technical artefacts their use-plan analysis together with their ICE-theory of function ascription raises questions. But before I enter into a discussion of these questions, let me comment on Houkes en Vermaas' claim that their ICE-theory can deal with the proper-accidental desideratum and in connection to that with the malfunction desideratum. That will induce me to put their interpretation of these desiderata into question; a more elaborate discussion of the desiderata for theories of technical functions will be postponed until the next chapter.

As we have seen, Houkes and Vermaas account for the distinction between proper and accidental function ascriptions by way of their distinction between proper and improper use-plans. Proper use-plans are use-plans that are socially approved within a community, whereas improper use-plans are socially disapproved. Proper-function ascriptions are defined as function ascriptions relative to proper use-plans, and improper-function ascriptions relative to improper use-plans. Designers acknowledged as expert designers within a community will typically develop socially acceptable use-plans, that is, use-plans that will become proper use-plans. This social grounding of the distinction between proper and accidental functions explains why proper functions are more stable than accidental functions: they are based on use-plans whose social acceptability within a community is more stable. According to Houkes and Vermaas this distinction between proper-function and improper-function ascriptions accommodates the proper-accidental function desideratum.

One may question, however, to what extent their distinction between proper and accidental functions matches the traditional distinction between these two kind of functions. One aspect of this traditional distinction is that an artefact X with some proper function φ (e.g., to drive screws) may be used *as* ψ-er (e.g., as a chisel), in which case X has or is ascribed the accidental function to ψ. It is not clear whether this use falls under the heading of a socially disapproved, that is, improper use-plan. The fact that the screwdriver X is not used in accordance with a socially approved use-plan does not imply that the new use-plan is socially disapproved (the pair of properties socially approved-socially disapproved are contrarily and not contradictorily opposed to each other). Suppose that the screwdriver is used effectively as a chisel for the first time ever by some creative user and that this use-plan for the screwdriver has not yet acquired the status of being (dis)approved socially. So this use is not improper, but it is not proper either.[43] On the traditional construal of the proper/accidental distinction this is clearly a situation of accidental function

[43] In earlier work, Houkes and Vermaas leave open the possibility that the distinction between proper and accidental functions is not exhaustive (Vermaas and Houkes 2003, p. 265, note 3); this could be an example of a function ascription that is neither proper nor accidental. I will assume that the distinction is exhaustive, but not crisp, that is, there is a fuzzy zone where it is difficult to decide whether an artefact has a proper or accidental function.

ascription, but that is a conclusion they cannot draw. Apparently the ICE-theory is not able to account for this aspect of the traditional distinction between proper and accidental functions. This aspect, however, appears to be part of their proper-accidental desideratum, given their characterization of accidental functions as functions that are ascribed only transiently or occasionally. If this aspect is not intended to be included in this desideratum, then in my opinion it needs reconsideration since it is not in accordance with the phenomenology of functional discourse.

The proper-accidental desideratum is closely linked to the malfunctioning desideratum which is generally considered to be a touchstone for any theory of technical functions. It is instructive first to have a look at the rationale behind this desideratum. Why should a theory of technical functions allow the ascription of a proper function to an artefact that is not capable of performing that function? Note that this desideratum only pertains to proper functions; it is not necessary that the theory allows the ascription of an accidental function to an object that is not capable of performing that accidental function. The reason behind this desideratum is, in my opinion, rather obvious: if the theory would not allow the ascription of its proper function to a broken TV-set (broken, not because it has been smashed to pieces but because of a broken on/off switch), then that technical artefact would not be TV-set at all, *given the assumption that an object is an instance of a technical kind just in case it has or may be ascribed the corresponding proper function.* So, if the theory would not allow ascription of proper functions to malfunctioning technical artefacts, it would not be possible even to speak of, for instance, a broken TV-set. It is this assumption about proper functions and technical kinds that creates a close tie between theories of (proper) technical functions and theories about technical kinds. Houkes and Vermaas' ICE-theory is not a theory about technical kinds and they do not elaborate on how proper functions, as defined by their theory, are related to technical kinds. However, by posing the malfunctioning desideratum the way they do, they implicitly appear to assume that ICE-proper functions are related to technical kinds. Moreover they appear to assume that explicitly when they remark with regard to a damaged TV-set that (p. 110): "There is, after all, a point at which a television set stops having its original function, and becomes a former television set". That point is reached when the TV-set is damaged beyond repair, that is, when on the ICE-theory of functions it can no longer be ascribed a proper function.[44]

With regard to the malfunctioning desideratum, the greatest stumbling blocks for the ICE-theory are known-to-be-broken artefacts. Houkes and Vermaas go to great length explaining how the ICE-theory of functions may satisfy the malfunctioning desideratum in those cases. They do so by introducing the distinction between artefacts having a capacity and exercising that capacity. That distinction forces them in turn to introduce new conditions on the function-ascribing agents because a parked car is not exercising its capacity, but it is surely not a case of malfunctioning (p. 108).

[44] See also p. 149, where they discuss the persistence conditions for artefacts. These run parallel to the conditions for ICE-function ascriptions: "A car that is wrecked beyond repair in a crash is then no longer a car, but a mere aggregate of twisted steel and plastic."

I will not enter into a discussion of whether these conditions are adequate or whether the distinction between having and exercising a capacity can be fruitfully upheld. In my opinion, a much more promising way out of this predicament is by questioning whether proper functions, as defined by the ICE-theory, define the technical kind of the artefact involved.

My main reason for drawing attention to the relation between ICE-proper functions and technical kinds is that I have great concerns about whether user-centred approaches to technical functions, such as the one of Houkes and Vermaas, but also Preston's, can do justice to function ascriptions and function-kinds ascriptions (in general and in engineering practice in particular) as long as it is assumed that proper functions, as defined by those user-centred approaches, determine technical kinds. Houkes and Vermaas start off from a phenomenology of artefact *use* and propose a definition of artefact functions in terms of *use*-plans. Admittedly, they bring in the physical structure of technical artefacts and the possible role of designers in developing socially approved use-plans, but finally they end up with a definition of proper functions in terms of approval by social groups of *users*. As long as it is assumed that proper functions determine technical kinds, this means that ultimately users determine what kind of artefact a thing is. Houkes and Vermaas are aware that it is a consequence of their plan-centred metaphysics of artefacts that the use-plan in which an object is embedded determines its technical kinds (p. 150):

> …an object is one artefact with respect to one use plan or community of users, and a different artefact with respect to another plan and community. There would be no single correct answer to the question what type of artefact an object is: one person's screwdriver is another's can opener.

Such a chameleonic metaphysics of artefacts is in my opinion too versatile to be in correspondence with the way we deal with function kind-ascriptions in daily life. The use approach creates, in my opinion, a blind spot for the role of designers and makers of technical artefacts in determining what kind of artefact an object is. I have no trouble with users playing a prominent role in defining proper functions.[45] However, as I will argue in the next chapter, proper functions thus defined have to be clearly distinguished from proper functions on the basis of which technical kinds are defined.[46]

For now, let me turn to another aspect of the ICE-function theory, namely whether and how it reflects the dual nature of technical artefacts. Houkes and Vermaas

[45] See also Hansson (2006) and Scheele (2005, 2006).

[46] Note that Houkes and Vermaas run up against the limits of their use-plan approach in reconstructing functional descriptions when discussing functional-role ascriptions by analysts (p. 99); these cannot be reconstructed as function ascriptions relative to use-plans and so they are conceptually different. Interestingly, functional-role ascriptions involve describing a component as functioning as a φ-er (one of the few times that the notion of a φ-er turns up in their analysis), that is, as an instance of a particular technical kind, whereas function ascriptions relative to use-plans involve the ascription of the capacity to φ as a function. This may be taken as an indication that issues about kindhood are conceptually different from proper functions defined relative to use-plans.

characterize their theory as a function-ascription theory and state that this makes their proposal an intentional theory of functions (p. 78):

> On our function theory, the designer ascribes these capacities as functions to the artefacts involved. This makes our proposal primarily an intentional function theory. Function ascriptions to artefacts are determined by the beliefs of the designers of the use plans for the artefacts. It explicitly characterizes justifiable *ascriptions* of functions by agents on the basis of their beliefs and actions, instead of defining functions as *properties* that artefacts have independently of beliefs and actions.

It is not clear what kind of function ascription, descriptive or performative, they have in mind here. It seems that a descriptive function ascription is out of the question since that would not make their theory an intentional function theory (see my discussion in section 3.4).[47] So, that leaves the option of a performative function ascription, that is, Houkes and Vermaas' function ascription amounts to a function assignment. On this interpretation the ICE-theory may be taken to be an attempt to state the conditions under which function assignments may be considered to be adequate or justified (not true, since function assignments are different from epistemic function ascriptions).[48] As such their ICE-function theory may be considered to be a theory of justified function assignments. Apart from the claim that their function theory is intentional because it is a theory of function ascriptions, Houkes and Vermaas claim that the ICE-function theory is an intentional theory of functions because of the I-condition. That, however, does not make much sense. The I-condition simply states the necessity of certain epistemic beliefs on the part of the assigning agent and these epistemic beliefs do not and cannot assign or confer any functional features on a technical artefact. So, it is not because of the I-condition that technical artefacts have a dual nature, but because of the function assignment to an object that is manipulated in the course of the execution of a use-plan. The ICE-theory may therefore be taken to be a proposal for the conditions under which a function assignment may be considered adequate or allowed. As regards the dual nature of technical artefacts, a similar kind of remark applies to the C-condition; it states an epistemic condition in which reference is made to physical features of the object to which a function is ascribed. That does not confer a dual nature on the object of the function ascription (assignment). It is this function assignment together with the physical features of the object of the function assignment, not the beliefs about these physical features, that confer a dual nature on technical artefacts.

In closing, let me comment briefly on Houkes and Vermaas' dual definition of technical artefacts quoted above, for it provides a nice stepping stone for looking

[47] However, Vermaas (2009) presents the theory as an epistemic theory, that is, as a descriptive function-ascription theory (since function assignment is not an epistemic act). Hansson (2006, p. 21) also interprets an earlier version of the ICE-theory in the descriptive sense.

[48] If the ICE-theory is taken to be a theory that spells out the conditions for justified function assignments, then the task remains of working out the details of mind-dependent epistemic and ontological theories of functions in which these function assignments play a role; see, for instance, the suggestion by Hansson (2006, p. 22).

ahead at what I will to do in the next chapter. The definition is intended to state necessary and sufficient conditions for an object x to be an artefact a of the type t. Surprisingly, the definition leaves open whether the act of making x or of using x determines the type or technical kind t of x. If I am right that Houkes and Vermaas tacitly assume that the ICE-proper functions determine technical kinds, then it follows that use, not making, determines technical kind. In their comments on the dual definition of artefacts they indeed suggest that the first condition may fix whether an object is an artefact or not and the second condition may determine the types (kinds) of artefacts, which means that use-plans determine technical kinds. So, users 'make' an object an artefact of a particular kind by ascribing proper functions to it, not its maker(s). In my opinion, the maker (not in the sense of who actually made a thing but in the sense of the inventor/creator) has as much if not more claims to authorship of artefacts kinds than the user does, which means that the maker determines what kind of object (s)he has created. In the following chapter I will propose a theory of technical functions that is intended to do more justice to the role of makers in determining the technical kind to which an artefact belongs.

3.6 Conclusion

I have distinguished between two functional properties, being for φ-ing and being a φ-er. Theories of technical functions so far have concentrated on the first property and have left the relation between the property of being for φ-ing and being a φ-er mostly in the dark. One of the main conclusions to be drawn from this chapter is that in order to clarify the relation between these two properties we need, alongside a theory of technical functions, also a theory of technical kinds. Furthermore I have discussed the general form of epistemic and ontological theories of functions. Following a suggestion by Hansson, I have pointed out the two different senses, descriptive and performative, in which the notion of function ascription may be used, and I have analysed the role of function assignments in mind-dependent theories of functions.

My discussion of theories of technical functions shows a diversity of approaches. Searle stresses the role of human intentions through the notion of assignment of functions; for him technical artefacts are mind-dependent objects. He acknowledges the role of physical features in case of technical artefacts, but it is not clear how these physical features constrain the assignment of functions and he does not discuss how the assignment of functions (being for φ-ing) is related to being an instance of a technical kind (being a φ-er). Preston treats proper functions of technical artefacts analogously to proper functions in biology; she takes them to be defined in terms of use and reproduction histories. Although human intentions may play an important role in the use and reproduction of technical artefacts, this approach downplays the role of human intentions with regard to proper functions; they are not assigned, in the way Searle claims, but the factual history of use and reproduction of its ancestors determines whether an artefact has a certain proper function or not.[49] The physical features of technical artefacts play only a secondary role in her approach since she allows phantom functions to be real proper functions. Houkes and Vermaas present an action-theoretic analysis of functions of technical artefacts; artefacts are ascribed functions relative to use-plans. They claim that their ICE-function theory clarifies the role of human intentions and of physical features in function ascriptions. I have argued that it is more plausible to interpret the ICE-function theory as an account of the conditions under which function assignments may be considered adequate or allowed. Furthermore, the way the ICE-theory deals with post-hoc malfunctioning appears rather contrived.

One common feature of the theories discussed is that they do not address explicitly the issue of the relation between (proper) functions and being an instance of a technical kind; by and large they simply assume that technical kinds are defined as (proper) functional kinds. In the following chapter I will question this assumption and argue that creators (designers) and users of technical artefacts may each in their own way assign proper functions to technical artefacts, but that the corresponding proper functions have to be carefully distinguished especially when it comes to the definition of technical kinds. Another common feature is a strong focus on a user approach to theories of technical functions, which, combined with the assumption about proper functions and technical kinds, leads to the rather remarkable view that users and not the inventors or makers of technical artefacts are the creators of new technical kinds.

References

Aristotle. *Physica*. The works of Aristotle, Vol. II, David Ross ed. (1970), Oxford: At the Clarendon Press.

Bigelow, John, and Robert Pargetter. 1987. Functions. *Journal of Philosophy* 84 (4):181–196.

Bloom, Paul. 1996. Intention, history, and artifact concepts. *Cognition* 60 (1):1–29.

Cummins, Robert. 1975. Functional analysis. *Journal of Philosophy* 72 (20):741–765.

Dipert, Randall R. 1993. *Artifacts, Art Works, and Agency*. Philadelphia: Temple University Press.

Duzi, Maire, Bjorn Jespersen, and Pavel Materna. 2010. *Procedural semantics for hyperintensional logic; foundations and applications of transparent intensional logic*. Springer.

Griffiths, Paul E. 1993. Functional analysis and proper functions. *British Journal for the Philosophy of Science* 44:409–422.

Hansson, Sven Ove. 2006. Defining technical function. *Studies in History and Philosophy of Science* 37:19–22.

Hilpinen, Risto. 1992. On artifacts and works of art. *Theoria* 7:58–82.

[49] In (Preston 2008, p. 28) she writes that she doubts whether "the proper functions of artifacts *are* dependent on human intentions in any relevant sense." In her latest proposal for the definition of proper functions (Preston 2009a), however, human intentions do appear to play a role because of the reference to the notion of explanation.

Houkes, Wybo, and Pieter E. Vermaas. 2010. *Technical functions: on the use and design of artefacts*. Philosophy of engineering and technology. Dordrecht: Springer.

Jespersen, Bjorn, and Massimiliano Carrara. 2011. Two conceptions of technical malfunction. *Theoria* 77:117–138.

Kitcher, Philip. 1998. Function and design. In *The philosophy of biology*, eds. David L. Hull, and Michael Ruse. Oxford readings in philosophy. Oxford: Oxford University Press.

Kroes, Peter. 2010a. Theories of technical functions: function ascriptions versus function assignments, Part 1. *Design Issues* 26 (3):62–69.

Kroes, Peter. 2010b. Theories of technical functions: function ascriptions versus function assignments, Part 2. *Design Issues* 26 (4):85–93.

Kroes, Peter, and Pieter Vermaas. 2008. Interesting differences between artifacts and natural objects. In *APA newsletter on philosophy and computers*.

Krohs, Ulrich. 2004. *Eine Theorie biologischer Theorieen*. Berlin: Springer Verlag.

Krohs, Ulrich. 2009. Functions as based on a concept of general design. *Synthese* 166 (1):69–89. doi:10.1007/s11229–007–9258–6.

McLaughlin, Peter. 2001. *What Functions Explain: Functional Explanation and Self-Reproducing Systems*. Cambridge: Cambridge University Press.

Millikan, Ruth. 1984. *Language, Thought and Other Biological Categories*. Cambridge, MA: MIT Press.

Millikan, Ruth. 1999. Wings, spoons, pills and quills: a pluralist theory of function. *Journal of Philosophy* 96 (4):192–206.

Neander, Karen. 1991. Functions as selected effects: the conceptual analyst's defense. *Philosophy of Science* 58 (2):168–184.

Perlman, Mark. 2004. The modern philosophical resurrection of teleology. *The Monist* 97 (1):3–51.

Preston, B. 2000. The function of things: a philosophical perspective on material culture. In *Matter, materiality and modern culture*, ed. Paul Graves-Brown, 22–49. London: Routledge.

Preston, B. 2006. Social context and artefact function - Discussion. *Studies in History and Philosophy of Science* 37 (1):37–41.

Preston, Beth. 1998. Why is a wing like a spoon? A pluralist theory of function. *Journal of Philosophy* 95 (5):215–254.

Preston, Beth. 2003. Of Marigold Beer: A Reply to Vermaas and Houkes. *British Journal for the Philosophy of Science* 54 (4):601–612.

Preston, Beth. 2008. The shrinkage factor: comment on Lynne Rudder Baker's "The shrinking difference between artifacts and natural objects". *APA newsletter on philosophy and computers* 8 (1):26–28.

Preston, Beth. 2009a. Biological and cultural proper functions in comparative perspective. In *Functions in biological and artificial worlds: comparative philosophical perspectives*, eds. Ulrich Krohs, and Peter Kroes, 37–50. Cambridge Mass.: MIT.

Preston, Beth. 2009b. Philosophical theories of artifact function. In *Handbook of philosophy of technology and engineering sciences*, ed. Anthonie Meijers, 213–234. Elsevier.

Quine, W. V. 1980. *From a logical point of view: 9 logico-philosophical essays*. 2nd ed. Cambridge, Mass.: Harvard University Press.

Scheele, Marcel. 2005. The proper use of artefacts: a philosophical theory of the social constitution of artefact functions. Delft University of Technology; PhD Thesis, Delft, The Netherlands.

Scheele, Marcel. 2006. Social aspects of technical functions: Proper function and proper use. *Studies in History and Philosophy of Science* 37 (1):23–36.

Searle, John. 1995. *The Construction of Social Reality*. London: Penguin Books.

Thomasson, Amie L. 2003. Realism and human kinds. *Philosophy and Phenomenological Research* 67 (3):580–609.

Thomasson, Amie L. 2007. Artifacts and human concepts. In *Creations of the mind: essays on artifacts and their representations*, eds. Stephen Laurence, and Eric Margolis, 52–73. Oxford: Oxford University Press.

Vermaas, Pieter E. 2009. On unification: taking technical functions as objective (and biological functions as subjective). In *Functions in biological and artificial worlds: comparative philosophical perspectives*, eds. Ulrich Krohs, and Peter Kroes, 69–87. Cambridge Mass.: MIT Press.

Vermaas, Pieter E., and Wybo N. Houkes. 2003. Ascribing functions to technical artefacts: A challenge to etiological accounts of functions. *British Journal for the Philosophy of Science* 54 (2):261–289.

Vermaas, Pieter E., and Wybo N. Houkes. 2006. Technical functions: A drawbridge between the intentional and structural natures of technical artefacts. *Studies in History and Philosophy of Science* 37 (1):5–18.

Wetzel, L. 2006. Types and tokens. In *The Stanford encyclopedia of philosophy (Summer 2006 Edition)*, ed. Edward N. Zalta. (editor): http://plato.stanford.edu/archives/sum2006/entries/types-tokens/.

Chapter 4
Proper functions and technical artefact kinds

4.1 Normativity and technical functions

One of the touch stones for assessing the viability of theories of technical functions we have come across time and again is how they account for malfunctioning artefacts. Malfunction statements are just one kind of statements with regard to technical artefacts that on the face of it are normative. They include evaluative statements such as:

– "This φ-er malfunctions, or is not capable to perform its function to φ":
  "This TV-set is broken."
– "This φ-er is a good/poor φ-er"
  "This knife is a good/poor knife."
– "*X* is good for φ-ing"
  "This object is good for driving screws."[1]

On top of evaluative statements we also make prescriptive ones of the following kind:

– "This φ-er ought to/is supposed to φ"
  "This TV-set ought to/is supposed to produce TV-images."

All these different kinds of statements, it seems, are normative in character. The evaluative statements refer (implicitly) to norms of goodness, whereas the prescriptive ones refer to norms of behaviour.

The above statements may look like normative ones, but are they really? If they are, then *prima facie* technical artefacts may have normative properties by

[1] Note that in statements of the form "*X* is good for φ-ing" *X* may also stand for objects without a function (such as natural objects).

P. Kroes, *Technical Artefacts: Creations of Mind and Matter: A Philosophy of Engineering Design*, Philosophy of Engineering and Technology 6,
DOI 10.1007/978-94-007-3940-6_4, © Springer Science+Business Media B.V. 2012

themselves. So, there would be 'norms in the artificial world'.[2] This would put a heavy burden on theories of technical functions for it would imply that these theories will have to account for these intrinsic normative features. To illustrate that the normative nature of the above statements may be disputed, take the malfunction statement that this TV-set malfunctions (is out of order). In one sense, this does not appear to be an evaluative statement at all, but a statement of an empirical fact; it is supported by the observation that when I turn on this TV-set it does not produce any images. When I bring the set to a repair shop and when after investigating it, the repair man comes to the conclusion that some part malfunctions, he is not making a normative statement about the part, but an empirical one (that may be defeated by facts about the TV-set). A sign of the empirical/factual nature of such statements is that they can be true or false (the TV-set may not malfunction after all since I may have forgotten to connect the TV-set to the power supply, and the repair man may mistakenly claim that a particular part malfunctioned). In another sense, this malfunction statement may be taken to be normative or closely related to a normative statement, if the statement "This TV-set malfunctions" is taken to mean or to imply that this TV-set ought to or is supposed to perform its function or is to be taken as a recommendation not to use this TV-set if one wants to watch TV.

One of the problems in dealing with the question whether the above statements are normative or not, is that there is no generally accepted interpretation of the notion of the normative. A widespread way of delineating the domain of the normative is by opposing it to the domain of the factual, but depending on how the notion of fact is interpreted, facts may include values (Raz 1990 (1975), p. 18). Moreover, moral realists defend the idea that there are moral facts and that moral judgments may be true or false. These views imply that the domain of the normative is not opposed to the domain of the factual. Fortunately, the clarification of the 'normative' character of the above statements that we are after here does not hinge on how the domain of the normative is characterized in a general way. What we need is a clarification of what is so peculiar about technical artefacts, as distinct from physical objects without a function, which makes it possible to make meaningful statements about them of the kinds presented above. Whatever factual statements can be made about a physical object such as a Helium atom, the claim that a Helium atom malfunctions or that a Helium atom is good as a Helium atom are not among them, for the simple reason that such claims do not make sense. However, they do make sense with regard to a TV-set or technical artefacts in general. Even if these normative statements about technical artefacts turn out to be factual on some general account of normativity, then there appear to be two kinds of factual statements about technical artefacts, one kind of which is similar to factual statements about physical objects and one kind that has no counterpart for physical objects. In that case, the latter kind concerns factual statements about what may be

[2] See (Davies 2001) for a discussion of the analogous idea that there are norms in nature in relation to normative statements about biological functions.

termed the 'goodness' of technical artefacts. What we need is a clarification of statements about the goodness of technical artefacts, be they factual or not on some general account of the normative.

One of the first to analyse in depth the meaning of statements about the goodness of technical artefacts is Von Wright (1963, Ch. 2–3).[3] He calls one form of goodness that may be attributed to technical artefacts (implements, instruments and tools) *instrumental* goodness, which is related to serving a purpose well. He distinguishes three ways in which technical artefacts may be called instrumentally good. First, a thing may be called good for a purpose ("this knife is good for cutting"). Second, a thing may be called good of its kind ("this knife is, as a knife, a good knife"). Third, a thing may be good as a knife without being a knife ("this thing may be used successfully as a knife"). From a logical point of view, the first way of attributing instrumental goodness is more basic than the second and the third (1963, p. 20):

> To attribute instrumental goodness to some thing is *primarily* to say of this thing that *it serves some purpose well*. An attribution of instrumental goodness *of its kind* to some thing presupposes that there exists some purpose which is, as I shall say, *essentially associated* with the kind and which this thing is thought to serve well. An attribution of instrumental goodness *of its kind* to some thing is thus *secondary* in the sense that it logically presupposes a judgment of goodness *for some purpose*.

The purpose essentially associated with the kind is closely related to what we have been calling the proper function. Another form of goodness discussed by Von Wright concerns expressions in which something is said to be good for a thing or being (e.g. lubrication is good for a car). Von Wright calls this utilitarian goodness or usefulness. Usefulness is primarily related to being good for a purpose, whereas instrumental goodness is related to serving a purpose well. In the following I will mainly concentrate on instrumental goodness.

So Von Wright distinguishes the following three ways of relating instrumental goodness and technical artefacts:

A. *X* is good for *P*	(*X* is good for cutting).
B. *X* is good as a *K*	(*X* is good as a knife, but is not a knife)
C. *X* is, as a *K*, a good *K*	(*X* is, as a knife, a good knife)

K refers to an artefact kind and *P* to a purpose. Let me briefly point out how these various statements are related to the two different ways in which functional properties may be attributed to an object (see section 3.2):

(1) "*X* is for φ-ing", and
(2) "*X* is a φ-er'.

[3] There are only a few in depth analyses of normative statements about technical artefacts. Here I concentrate on the analyses by Von Wright (1963) and Franssen (2006, 2009b, 2009a). See also Vaesen (2008). For an analysis of some of the logical relationships between value statements about functionally specified categories, among which technical artefacts, and their subcategories, see Hansson (2006a) and Franssen (forthcoming).

It is rather obvious to take A as the counterpart of:

(1') "*X* is good for φ-ing",

and C as the counterpart of:

(2') "*X* is a good φ-er'.

In my opinion, B may be interpreted as a special case of A in the following way. Take *P* to be the purpose essentially associated with kind *K*. Now, if an object *X* is good as a *K*, then it can be used with advantage for the purpose essentially associated with the kind *K*, that is, it can be used for the purpose *P*. That is precisely how Von Wright interprets A (1963, p. 43). The only difference between A and B is that B implicitly implies that *X* is not a member of the kind *K*. This means that B is a special case of the more general case A, which leaves open whether *X* is a member of the kind *K* which has purpose *P* essentially associated with it (if such a kind exists at all). I will therefore assume that (1') covers the two cases A and B.

Von Wright considers statements of the kinds A to C to be *value*-judgments. But they are a particular kind of value judgments since they may be 'objectively' true or false (Von Wright 1963, p. 29 and 48–49). To illustrate this he analyses the example of the sharpness of a knife. Suppose somebody wants to cut something smoothly and on that ground prefers a sharper knife to a lesser sharp one. What makes one knife better than another in this situation is the sharpness of its knife-edge. This better-making property, the sharpness of the knife-edge, is *causally* related to the smoothness of the cut. The smoothness of the cut in turn is *logically* related to the betterness of knives, given the subjective setting of a purpose (to make smooth cuts). Thus sharpness is causally related to the goodness (betterness) of a knife by being causally related to smoothness which itself is logically related to the goodness (betterness) of a knife (within the context of a subjective goal). Since sharpness is in this way *causally* related to the goodness of knives, value judgments about the instrumental goodness of knives may be true or false. In short, given that I want to use a knife to realize a certain end, it is a matter of fact whether one knife realizes the end better than another one.

Although value judgments of instrumental goodness may be true or false, Von Wright refrains from calling them descriptive judgments (1963, p. 30), since they are not primarily used for the purpose of describing. They nevertheless have descriptive content, on the basis of which they may be true or false. One of the most important uses of calling something good is according to Von Wright not descriptive but *commending* the use of the thing. If a person commends the use of a thing by calling it good or better, then she states that under certain conditions one has a *reason* to use that thing. These conditions are that one is in pursuit of the purpose with regard to which that thing is called good and that one prefers the better thing when there is the possibility of a choice of various things to use.

What conclusions may be drawn from Von Wright's analysis with regard to the normative status of the statements with which we started this section? His interpretation of value judgments of instrumental goodness makes it difficult, if one adheres to the traditional opposition between normative and factual judgments, to classify these statements unambiguously as factual or normative. Von Wright explicitly characterizes judgments of instrumental goodness as evaluative and not as descriptive judgments. He remarks that the descriptive content of these evaluative judgments may be taken to be the meaning or sense of these judgments. The descriptive content of such judgments is not empty since "genuine judgments of instrumental goodness are always objectively true or false judgments" (1963, p. 29). Such judgments are objectively true or false because they express causal connections in the world. In this respect they are not different from any judgment about causal relationships, for instance, those made in the physical or engineering sciences. But this means that these evaluative judgments, *in so far as their meaning is concerned*, are very similar to factual or empirical statements. In other words, the *prima facie* evaluative statements that we started with are, in so far as their meaning is concerned, not normative statements. They do not imply that the objects involved have (intrinsic) normative properties. This is in line with our earlier remark that the observation that a technical artefact malfunctions often expresses simply an empirical fact.

The foregoing does not exclude that these evaluative judgments may have normative content when they are used in a recommendation mode. Then they commend (or not) the use of *X* under certain circumstances, which amounts to stating reasons for using *X*. According to Von Wright such recommendations always contain a conjectural element, in the sense that calling something a good knife involves expectations about future behaviour of the knife (Von Wright 1963, p. 27). These expectations make it possible to make a direct link between judgments involving instrumental goodness and 'ought to/is supposed to' statements. The statement that *X* is a good knife states reasons for expecting a certain kind of behaviour when used, and this may be expressed by saying that *X* ought to or is supposed to show that behaviour when used.[4] Also malfunction statements may be interpreted in their recommendation mode as stating reasons for not using the objects involved.

Interpreted in this way, the *prima facie* evaluative statements at the start of this section are normative in stating reasons for prospective users to use or not to use *X*; these reasons justify (up to a certain extent) expectations about behaviour of *X* which are typically rendered in the form of '*X* ought to…' or '*X* is supposed to…' statements. So the evaluative and prescriptive judgments appear closely related.

It is also to be noted that the statement '*X* is a knife', which does not explicitly refer to instrumental goodness and on the face of it is not an evaluative statement, may be taken as recommending use. It may be taken to state a reason for using *X* when one is in need of a knife. In other words, an agent may be taken to have reasoned expectations about the future behaviour of *X*; *X* ought to or is supposed to behave in a certain way. Thus, ought to/is supposed to statements may also be related

[4] I will not enter into a discussion of whether the descriptive content of the judgment "X is a good knife" has to be true or not in order for it to imply a reason for use. This involves the distinction between objective and subjective reasons; see, for instance, Dancy (2000). It appears that Von Wright allows that reasons are subjective; see (1963, p. 32).

to mere function-kind attributions.[5] If indeed statements of the form "*X* is a φ-er" imply ought to/is supposed to statements, then it follows that statements of the form "*X* is a good/poor φ-er" are not normative statements simply because they make use of the seemingly evaluative terms 'good' or 'poor'.

The upshot of the foregoing is that the evaluative and prescriptive statements about *X* as well as mere function attribution to *X* do not make *X* an entity that is intrinsically normative in the sense that *X* in itself is good or bad (poor) or that it of itself ought to behave in a certain way. These statements are normative in so far as they are used in the recommendation mode, which means that they are normative in relation to potential users of *X*; they state that these potential users have reasons to use or not to use *X* because they may expect certain behaviour of *X*.

Franssen (2006, 2009a) has presented an analysis of normative and evaluative statements about technical artefacts that leads to more or less the same conclusion as Von Wright's analysis, namely that there is nothing intrinsically normative about technical artefacts. He goes beyond Von Wright's analysis by, among other things, explicitly analysing 'ought to/is supposed to' statements and by taking into account the distinction between artefact types and tokens.

Starting from Dancy's characterization of the normative in terms of the difference that facts about the world make to what we believe, desire and do, evaluative statements of the form "*x* is a good K" with K denoting an artefact of a particular functional kind are analysed by Franssen in the following way (2006, p. 46):

> '*x* is a good K' expresses the normative fact that *x* has certain features *f* and that because of these features, if a person *p* wishes to achieve the result of K-ing, then *p* has a reason to use *x* for K-ing.

Franssen assumes that K-ing is the proper function of *x*, that is, the function for which it is designed. This means that in our terminology '*x* is a good K' stands for '*x* is a good φ-er'. Similarly, malfunction statements are interpreted in terms of reasons for persons not to use the malfunctioning object. Thus, evaluative statements about technical artefacts state normative facts in terms of reasons for people to do something (or not) with those artefacts. The normativity of these judgments is therefore grounded in human intentionality, in the purposes they pursue.

In a similar vein, prescriptive statements related to functions, such as "Object *x* is supposed to φ" or "Object *x* ought to φ" are interpreted. They are not to be taken literally in the sense that object *x* has (conclusive) reasons to behave in a certain way. These statements are prescriptive in form only, not in content. They are to be interpreted in terms of expectations of human agents with regard to object *x* (Franssen 2006, p. 54):

> When a person *p* says '*x* ought (not) to do K in circumstances *c*', *p* expresses the opinion that *p* is justified in expecting that *x* will (not) do K in circumstances *c*.

[5] For more details about how to interpret function ascriptions as normative facts, see (Franssen 2006, p. 52–53); he argues that function ascriptions express normative facts of a theoretical kind (i.e., reasons about what to believe) instead of normative facts of a practical kind (i.e., reasons about what to do). For a criticism of Franssen's analysis on this point, see Dancy (2006, p. 59).

Franssen points out two different grounds for the justification of these expectations, namely rational or epistemic and moral considerations. Epistemically justified 'ought to' statements are not specific for objects with a technical function; also with regard to natural objects, for instance, an electron, such 'ought to' statements can make sense ("This electron ought to show such and so behaviour"). By contrast, morally justified 'ought to' statements only apply to artefacts (and not to physical objects or biological items), because they express a certain right on the basis of for instance a (implicit) promise.[6] Consider the case of a car owner saying "This car ought to start now", after having paid a big repair bill to her car mechanic; with this statement she expresses a moral right on her part of the car functioning well. As in the case of evaluative statements, 'ought to/is supposed to' statements derive their normativity from human intentionality and make no sense beyond the domain of human behaviour.

So far we have considered only statements about individual technical artefacts being good or poor. In order to deal with more general normative statements about technical artefacts Franssen introduces the notions of an artefact *token*, artefact *type* and an artefact *kind* (2006, p. 48). An artefact token is a particular technical artefact, for instance 'this knife over here'. Such an artefact is a member of the artefact kind 'knife'. An artefact kind is described as an artefact defined in terms of its functional role and the fact that it is designed for performing this functional role (depending on the various kinds of functions a knife may perform, sub-kinds of knifes may be distinguished). Apart from artefact kinds Franssen introduces the notion of an artefact type, which is an artefact as defined by its functional role and by its total design/manufacture history (for instance, an eight-inch chef's knife from the firm Zwilling J.A. Henckels).[7] This means that for an artefact type, there are always a specific number of tokens actually made at any particular time. So not the mere fact that an object is designed and performs a function, but also its particular design and manufacturing features matter in identifying the artefact type of which it is a token.[8]

On the basis of these distinctions Franssen argues that not only artefact tokens may be qualified as good or poor, but that the same applies to artefact types. Calling an artefact type good means that it is based on a good design and manufactured in a good way. This means that new tokens of this artefact type may be expected to be good tokens compared to tokens of a different artefact type that may be used for the same purpose.[9] Contrary to artefact types, it does not make sense to call artefact

[6] In certain circumstances morally justified statements may, in my opinion, also be made with regard to natural (biological) objects; see the example of the shells in our discussion of Preston's theory of functions in section 3.5.2.

[7] What Franssen calls 'artefact kinds', I have been referring to so far as 'technical kinds' and what he calls 'artefact types' I will be referring to later as 'technical artefact kinds'. In later work Franssen(2009a) rephrases his distinction as a contrast between functional kinds and artefact kinds.

[8] According to Franssen the level of detail required for identifying artefact types is context dependent.

[9] 'New' tokens in order to exclude the poor functioning of tokens due to wear and tear.

kinds good or poor since "goodness and poorness discriminate performance within a particular functional domain and therefore do not apply to the functional domain as such" (2006, p. 49). Whereas artefact types may be evaluated as good or poor, they cannot be said to malfunction since that would mean that all tokens of the type are unable to perform the function of the type; in that case the artefact type can hardly be said to belong to that functional kind.[10]

Looking back on Von Wright's and Franssen's analyses the first and main conclusion to be drawn is that the normativity associated with technical artefacts, in particular with their functions, is not intrinsic to them, but derives from the normativity of intentional human action.[11] For theories of technical function this means that these theories do not have to provide an account for the normativity of functions taken as a kind of normativity *sui generis*, apart from normativity related to human action. The foregoing does not mean that theories of technical functions do not have to account for the normativity of functions. On the contrary, they have to; the normativity of functions remains one of the main criteria for assessing the viability of theories of functions. Any acceptable theory of technical functions should allow for normative statements about technical artefacts, in particular, should allow for malfunction statements. However, the account of normativity must clarify how the normativity of functions is grounded in the normativity of human action. This outcome of our analysis of the normativity of technical artefacts is very much in line with our dual nature thesis. In order to explain how technical artefacts perform their function, we have to make reference to their physical features, but normative statements about the functioning of technical artefacts cannot be explicated in terms of these physical features alone; for that, reference to human intentions and the normativity involved in human action is necessary.

Franssen also addresses the question of how well existing theories of functions are able to account for the normativity of functions. He claims that the conclusion that there is no intrinsic normativity of functions in technology (nor in biology) has important consequences for function theories, in so far as normativity has up till now commonly been treated as a touchstone for those theories. He distinguishes between two rival types of theories of function, the causal-role and the etiological theories; to the latter belong Millikan's proper function theory and Neander's selected effect theory (Preston's theory also falls into this type). Causal-role theories have traditionally been charged of being unable to account for the normativity of functions, in particular for malfunction, whereas it was considered to be one of the great advantages of etiological theories that they could do so. Franssen denies that etiological theories have such an advantage. Any claim that only etiological

[10] McLaughlin (2009) argues that one way normativity enters function ascriptions is by the use of the type-token distinction; he claims that that distinction itself is normative since tokens can instantiate a type in better or worse ways. If the type-token distinction is taken to run parallel to the kind-instance distinction then this kind of normativity is not specific for technical artefacts.

[11] Davies (2001, p. 194, 214) reaches a similar conclusion; the only norms involved in functions are, according to him, epistemic in character; objects with a proper function "are not the bearers of norms of any sort."

theories are able to account for the intrinsic normativity of technical artefacts (and are thus to be preferred above causal-role theories) is ill founded for there is nothing intrinsically normative to account for.[12] He also points out that defenders of causal-role theories may account for malfunctioning by an appeal to the type-token distinction; a token of an artefact kind may be taken as malfunctioning when it does not show the behaviour that tokens of this type typically may be expected to show. This does not allow the attribution of a function to a malfunctioning object, yet proponents of causal-role theories might consider this interpretation to be good enough.[13] This, according to Franssen, restores the parity between causal-role and etiological theories of function regarding the issue of normativity of functions. Therefore, he (2009b, p. 121) concludes that "the advantage that the [etiological] theories have always claimed over their rival theories – of uniquely being able to account for malfunction – is illusory." It is illusory because there is no intrinsic normativity associated with technical functions which has to be accounted for.

A second conclusion to be drawn from our analysis of the normativity of functions is that it underscores the need to carefully distinguish between descriptive and performative aspects of goodness statements about technical artefacts. The statement that *X* is a good/poor φ-er contains the function ascription statement that *X* is a φ-er. We have seen that function ascriptions in general may be taken in a descriptive and a performative sense. This distinction also holds for evaluative function ascriptions. Following Von Wright, it may be argued that in its descriptive sense the statement that *X* is a good φ-er amounts to a non-normative, factual statement about causal connections in the world. In its performative sense this statement may be taken as a recommendation, as a statement giving reasons for agents to use or not use the artefact in particular circumstances. In this performative sense, the statement is normative.

In closing this section let me point out an interesting connection between the issue of the normativity of functions and the topic to which I turn next, namely technical kinds and technical artefact kinds. Consider the following three statements[14]:

"Object *X* is a screwdriver."
"Object *X* is a poor screwdriver."
"Object *X* is not a screwdriver."

Suppose that the object *X* referred to in these statements is physically one and the same object that undergoes only some changes in geometrical shape during its lifetime. Initially, *X* may be a (good) screwdriver, but because of wear and tear it may turn into a poor screwdriver. It may even become damaged to such a degree, that it

---

[12] It seems that *naturalistic* etiological theories à la Millikan can do so only on pain of committing some form of naturalistic fallacy (Franssen 2009b). For similar criticism on etiological theories, see Davies (2001, especially chapter 7).

[13] See also Davies (2001, p. 175 ff; 212) who claims that systemic malfunctions are not possible (because the malfunctioning object, lacking the function, is no longer a member of the relevant functional type), but he explains our inclination to attribute malfunction in such cases in terms of our expectations based on our experiences with well-functioning tokens of the relevant functional type.

[14] See also (Kroes 2003).

would no longer be considered, as a matter of objective fact, a screwdriver. So we may imagine a continuous spectrum of states of the physical object involved described variously as 'being a screwdriver' at one end, through 'being a poor screwdriver' to 'not being a screwdriver at all' at the other end of the spectrum.

This raises the issue of whether it is always possible to separate questions about whether an object is an instance of the technical kind screwdriver from questions about whether the object is a good or a poor screwdriver. In some situations the ascription of the property of being a member of a technical kind may be taken in a descriptive sense and justified on objective grounds (see, for instance, the quote from Searle in section 3.5.1.2 about an object being a screwdriver without any metaphysical doubt). In other situations the ascription may be taken more in a performative sense (as a recommendation to use an object for driving screws). A similar ambiguity occurs with regard to statements about the goodness of a screwdriver; they may also be taken in a descriptive and a performative sense. There are, however, also all kinds of intermediary cases in which it is not so clear whether descriptive or performative senses of function ascription and evaluative statements are intended. In those cases it may become difficult to separate issues about whether something is an instance of a technical kind from evaluative issues. Take again the example of the screwdriver. For establishing whether some *X* is a poor screwdriver or not it might be proposed to follow the procedure of first deciding whether *X* is a screwdriver or not, and if so, then adding an evaluative judgment about the quality of the object as a screwdriver. According to this line of thought, issues about being an instance of a technical kind can always be dealt with independently of evaluative issues. In my opinion this view is rather problematic. Issues about being an instance of a technical kind and evaluative issues may become inextricably intertwined. In the following sections I will leave such borderline cases aside and will focus primarily on situations in which normative issues are not considered relevant for issues about being an instance of a technical kind.

4.2 Function theories and technical kinds

At several occasions I have pointed out that in the function literature it is almost without exception assumed implicitly or explicitly that (proper) functions as defined in those theories define technical kinds. An object is an instance of a technical kind just in case it may be ascribed the (proper) function associated with that kind. In this section I argue that this assumption causes serious troubles for causal-role and evolutionary (etiological) function theories. For causal-role theories it makes it problematic to deal with malfunction and for evolutionary theories it leads to a circularity problem. Note that in the following I will make a distinction between the notions of technical kinds and of technical artefact kinds. Technical kinds are functional kinds; being an instance of a technical kind is determined by having the appropriate proper function. That is the notion that I have been using mainly up till now. By taking into account their physical and design features the instances of a technical kind may be subdivided into various technical artefact kinds. Technical artefact kinds are therefore defined in terms of functional and structural features.

About causal-role theories I can be brief, since we have already come across the malfunction problem several times. Because of the technical-kinds assumption causal-role theories face a problem with malfunction, similar to the one we discussed in relation to the ICE-theory (see section 3.5.3). According to causal-role theories a malfunctioning technical artefact cannot be ascribed its function because it lacks the relevant capacity. Together with the technical-kinds assumption this means that the object involved is not an instance of the corresponding technical kind. So it would be impossible to say that some object is a broken TV-set.[15]

At first sight, the technical-kinds assumption does not appear troublesome for evolutionary accounts of proper functions, such as Beth Preston's. These accounts allow for the ascription of a proper function even to a malfunctioning instance of a technical kind. However, for another reason these theories cannot define technical kinds with the help of proper functions as defined by those theories. This reason has to do with the fact that most, if not all, etiological theories make use of the distinction between types and tokens. The use of the notion of type is necessary to create a link between a (possibly) malfunctioning object and its predecessors, where the object and its predecessors are all taken to be tokens of the same type. Assuming that for the present purposes the type-token distinction may be taken to be equivalent to the kind-instance distinction, the technical kinds (types) that figure in these theories cannot be defined with the help of their proper function.[16] Otherwise they immediately run into a vicious circle as pointed out by Franssen (2009b, p. 106):

> Care must be taken to distinguish between the type that an item is presumed to be a token of, in receiving a function on account of a particular theory of function, and the functional type itself, that is, the type that is defined as consisting of all tokens that have this particular function. The type to which the theory of function refers cannot be the functional type, on pains of circularity: in order to know whether x is a token of the functional type X_F, it must be known whether x has the function F, but in order to know whether x has the function F, it must be known whether x belongs to the functional type X_F.

Thus, function theories that rely on the kind-instance (type-token) distinction cannot define technical kinds with the help of proper functions as defined by those theories.[17]

---

[15] See also Davies's remarks about the impossibility of systemic malfunctions (2001, p. 212); he does not see this as a decisive arguments against the theory of systemic functions.

[16] Intuitively, the type-token and kind-instance distinctions express the same difference, namely that between a general sort of thing and its particular concrete instances. In fact, the two notions appear to be used indiscriminately by some function theorists (see for instance Davies (2001, 212–3), who equates functional kinds and functional types). There may be general metaphysical/ontological reasons for distinguishing types from kinds (see (Wetzel 2006)), but I see no reason that they make it necessary to distinguish in the present context between artefact types and artefact kinds.

[17] Also Enç's analysis of the attribution of functions appears to suffer from the kind of circularity discussed here; he presents a definition of the attribution of a function to X, where X is a member of a natural kind and part of the identity conditions for being an X involve the function of X; see (Enç 1979).

So, after all, defenders of causal-role and etiological theories of proper function appear to share the same predicament. Albeit for different reasons, they have to come up with a definition or characterization of technical kinds that does not rely crucially on the notion of (proper) function as defined in these theories. Opinions differ on whether this is possible. Within the context of systemic (causal-role) function theories Davies apparently sees no possibility (2001, p. 212): "On what grounds other than possession of the defining capacity might they retain their membership in the functional type?" Franssen, however, suggests that technical artefact kinds may be characterized in a (proper-) function-independent way (2009b, p. 106–107):

> The predominant type concept in technology [...] seems to be the functional type: pump, knife, and so forth. It is not commonly recognized in technology that, apart from the functional type, another type is presupposed, which is identified by the physical and historical features of existing tokens of a functional type and by the design specifications associated with the type.

According to this suggestion, the properties on the basis of which an object can be identified as an instance of a technical artefact kind are not functional, but physical, historical and the properties described in the design specifications. However, this leads to the question how many physical, historical or design features of existing tokens of a functional type have to be included in the definition of a technical artefact kind. Inclusion of too many details may lead to an almost unrestrained proliferation of technical artefact kinds.

Here we run into the problem of what are appropriate individuation criteria for technical artefact kinds. We may use mainly functional criteria and only include some very general criteria concerning their physical make-up (for instance, that they perform their function in a mechanical way); that leads to technical artefact kinds being almost identical to functional (technical) kinds; any technical artefact with the appropriate function, satisfying very general physical criteria, is then an instance of that technical artefact kind. We may also use very fine-grained criteria for their physical make-up, for instance we may take into account minor changes in some of the materials of which the technical artefact is made in spite of the fact that they do not in any way affect the performance of its function. Given the inherent vagueness of the notion of technical artefact kind,[18] and given that pragmatic considerations usually play an important role in individuating technical artefact kinds, any attempt to come up with generally valid, unambiguous individuation criteria appears to be in vain.[19]

---

[18] See Simons (1995) who remarks that the identity conditions for (technical) artefacts are vaguer and more convention-bound than those for natural objects; see also (Thomasson 2003, p. 598–599).

[19] In my opinion, a promising way for individuating technical artefact kinds in terms of their function (input-output relations), the conditions under which this input-output relation obtains (system of interactions) and their structural and design features (object structure) has been presented by Soavi (2009). Pragmatic considerations may be taken to be part of the system of interactions.

In my opinion the malfunction problem for causal-role theories or the circularity objection against evolutionary theories can be solved only by defining technical kinds and technical artefact kinds in a way that is independent of the notion of (proper) function as defined by those theories. This means that apart from a theory of technical functions we have to come up with a theory of technical (artefact) kinds, *under the constraint* that in so far such a theory defines technical (artefact) kinds in terms of functional features the notion of function employed has to be independent of the notion of function as defined by the theory of function. In the following I will focus on technical artefact kinds and propose a theory that satisfies this constraint.

4.3 Technical artefact kinds

The starting point for my theory of technical artefact kinds is the interpretation of artefacts and artefact kinds developed by Thomasson (2003, 2007). It is an interpretation that is very much in line with our dual-nature conception of technical artefacts. She stresses that the intentions of humans making an artefact are constitutive for that artefact being an instance of a certain artefact kind; this means that technical artefacts are mind-dependent objects. However, not only the intentions of the makers are relevant; the intentions of the makers have to be successfully realized, which brings in the physical structure of artefacts. This means that both intentions and physical structure are involved in being an artefact of a certain kind.[20]

According to Thomasson artefact kinds are different from natural kinds. They are different from natural kinds because "the metaphysical natures of artifactual kinds are *constituted by* the concepts and intentions of makers, a feature that sets them crucially apart from natural kinds" (2007, p. 53). She rejects the idea that the nature of artefact kinds is determined by mind-independent properties "such as qualitative make-up, proper function, and historically proper placement" (2007, p. 53), a position defended, for instance, by Elder (2007). Artefact kinds have no "internal essences" shared by all the members of the artefact kind; instead what determines membership of the artefact kind is that they are things intentionally made, more specifically that they are made with the intention to create something of

[20] Because of their mind-dependence the ontological status of technical artefacts and of technical artefact kinds has been put into question; see for instance Baker (2008) and Kroes and Vermaas (2008). I will not enter here in a discussion of the general issue of how technical artefacts fit into the ontological structure of the world and whether they can be considered to be ontologically on a par with for instance natural kinds or not. Given my assumption about the basic ontology consisting of physical objects and intentions (see section 2.6), my aim in the following is to analyse what kind of physical objects and whose and what kind of intentions are involved in an object being an instance of a technical artefact kind. I will not address issues about what kinds of ontological dependence relations are involved between these physical objects and intentions and being an instance of a technical artefact kind; see for instance (Houkes and Meijers 2006; Meijers 2001; Kroes 2009).

that kind (for example, for something to be a chair it must have been made with the intention to make a chair). This means that artificial concepts exhibit self-referentiality. This self-referentiality seems to introduce a kind of circularity into the analysis, since the explication of what it means to be a hammer refers to the idea of a hammer. If we assume with Thomasson that the kind of properties that ground the reference of artefact terms is partly intentional in nature, this self-referentiality appears unavoidable. The making of the first instance of a (new) kind of artefact implies having a correct substantive idea of what is means to be an artefact of that kind. There is no real circularity involved here, however, since it is the prerogative of the creator of the first instance of an artefact kind to define what it means to be an instance of that artefact kind (see also Thomasson (2003) and Searle (1995, p. 32 ff)). Once the creator of the first instance has done that, it should be possible to describe roughly what a K is.

So, the sort of properties that ground the reference of artefact kind terms is intentional properties; whether something is a member of an artefact kind, as opposed to, for instance, a physical kind, is determined by intentional properties. Thomasson clarifies the intentions that are relevant for making an instance of an artefact kind K in the following way. When creating an artefact the maker's intention cannot always be described or made transparent in terms of making a new item of an already existing artefact kind K. That would exclude the possibility of making an instance of a new artefact kind. Moreover, just pointing at instances of an artefact kind and saying that one is making one of these will not do, since it is not clear what features of these instances are relevant for being a member of the kind K. This means that the intentions of the maker of a K should at least include what she calls a substantively correct concept of what a K is (2007, p. 59):

> the relevant sort of intention to make a thing of artifactual kind K must thus involve a substantive (and substantively correct) concept of what a K is, including an understanding of what sorts of properties are K-relevant and an intention to realize many of them in the object created [...] For a member of any essentially artifactual kind K to be created, it is also necessary that that intention be at least largely successfully realized.

When making an artefact, grand intentions, even when based on a substantively correct concept, are not sufficient; they have to be executed successfully to a large extent.

The foregoing means that when somebody is genuinely making an instance of the artefact kind K, (s)he must have a largely correct idea of what it means to be a K, which implies that there is some inherent vagueness in the concept of an artefact kind. Moreover, this largely correct idea has to be executed successfully to a large degree. The largely correct idea determines which features are relevant for being member of the artefact kind K. This does not mean that the criteria for membership of K are fixed once and for all by the substantively correct concept of a K of the person who made the first instance of a K (and through that created the artefact kind K). That would go against the fact that artefact kinds "are notoriously malleable and historical in nature" (2007, p. 62). Artefact kinds are historical kinds in the double sense that they come into existence by specific historical events (involving intentions) and that the criteria for membership of an artefact kind may change in the course of time. This change in membership criteria over time may occur because once a new artefact kind K has been introduced, subsequent makers of K's only need to have a *largely* correct idea of a K.

Thomasson's theory covers artefacts in general and leaves open the details of how it is to be applied to *technical* artefacts. In the following I put forward a proposal for how these details may be filled in. This means that I will have to explicate what the expression "a largely correct substantive idea" means in case of technical artefacts. For a start, let me paraphrase Thomasson's definition of being an instance of an artefact kind for being an instance of a technical artefact kind:

An object X is an instance of the technical artefact kind K iff X is the result of a largely successful execution of a largely correct substantive idea of a K.

At first sight, this definition of technical artefact (kinds) does not refer to functions at all. Nevertheless it does, but to see why and to what kind of functions, we have to unpack the notion of a largely correct substantive idea of a K. Thomasson remarks that her conception of artefacts and artefact kinds leaves open the possibility that intended functional features play a role in the largely correct substantive concept of an artefact kind K (Thomasson 2007, p. 59–60) and may therefore partly determine the boundaries of an artefact kind K. With regard to *technical* artefact kinds, however, the situation is different. For a technical artefact kind both intended functional features and intended structural (design) features are *necessary* ingredients of the largely correct substantive idea. To see why, suppose the substantive idea would consist only of functional features. In that case it would hardly make sense to call such a substantive idea a largely correct idea of a technical artefact kind.[21] In what sense may the purely input-output substantive idea of the nail clipper of chapter 2 be called a correct substantive idea of that kind of technical artefact? There is even no way this idea can be executed largely successfully. Similar problems may be raised with respect to a substantive idea that contains only a purely structural description of the nail clipper and leaves out all functional features. Also in this case it may be put into question whether the idea of a largely successful execution makes sense, because a structural description does not offer any criteria for which properties are relevant in making an artefact of that kind (this means that the structural description should fix every feature of the physical object that is *conceivably* relevant for performing the intended function). But even if such a substantive idea could be executed successfully, a more fundamental problem remains. As I have argued extensively in chapter 2, a substantive idea of an object that only refers to its structural features cannot be a substantive idea of that object as an instance of a technical artefact kind.

[21] Such a substantive idea may be taken to be a correct idea of what we have been calling technical kinds, since they are functional kinds.

There is yet another reason why a largely correct substantive idea of a technical artefact kind has to refer to functional and structural features. This has to do with the fact that part of the correctness of a substantive idea of a technical artefact kind resides in the fact that objects of that kind will be able to realize their functional features on the basis of their structural ones. A largely correct substantive idea of a technical artefact kind must take account of the fact that function and structure of technical artefacts are intimately related. Because technical functions are realizable in multiple ways, technical artefacts with the same (intended) function may belong to different technical artefact kinds because they are based on different *designs*. Roughly, a design shows or describes how the various parts of a technical artefact, each of which performs a sub-function, have to be arranged so as to be able to realize the overall function of the technical artefact.[22] With these different technical artefact kinds correspond different largely correct substantive ideas. So, the largely correct substantive idea has to contain a largely correct idea of the design which in the course of the execution of the substantive idea is embodied in matter. Of course, this raises problems about identity conditions for largely correct substantive ideas and thus about identity conditions for kinds of technical artefacts. Here we touch again upon the inherent vagueness of the notion of artefact kind. Whether two technical artefacts are considered to belong to the same artefact kind or not may be highly context-sensitive and therefore this problem cannot be solved in general.[23]

With the help of Hilpinen's (1992) characterization of artefacts it is possible to further explicate the role of functional features in Thomasson's notion of a largely correct substantive idea in the case we are dealing with technical artefact kinds. He considers an object to be an artefact only in case it is intentionally made by an agent under some description.[24] At least one of these descriptions must be a type-description (or sortal description), which fixes the identity of the artefact, that is, the technical kind to which it belongs; it determines the criteria on the basis of which the artefact may be distinguished from other kinds of artefacts (1992, p. 61). According to Hilpinen (1992, p. 67):

> Artifacts carry along certain type-descriptions in terms of which they are identified. As was mentioned earlier, such descriptions are normally associated with the function the artifact serves in the culture.

---

[22] I will have more to say on what a design is in the next chapter.

[23] To illustrate the complexity of defining artefact kinds in engineering practice, consider the Form, Fit and Function principle (see for instance http://www.dmsms.org/file.jsp?storename=Form_Fit___Function_Fundamentals.pdf). Artefacts with the same form, fit and function are supposed to be interchangeable. With regard to very simple components, such as nuts and bolts, the form, fit and function principle usually implies that those interchangeable artefacts are taken to belong to the same artefact kind. However, for more complex components, this is not the case; even when two artefacts have the same form, fit and function, they may be based on different designs and therefore be considered to be instances of different artefact kinds.

[24] The condition 'under some description' implies that a technical artefact cannot be produced by simply copying, in every minute detail, an existing instance of an artefact kind; the resulting object, that is a structural copy of an existing artefact, is not itself an artefact because it lacks the relevant description.

For *technical* artefacts the intended functional features play not just normally a role, but play an indispensable role in these type-descriptions, since they co-determine, together with structural features, the identity of technical artefacts. So, we may conclude that with regard to technical artefact kinds the function hidden in Thomasson's notion of a largely correct substantive idea is the *intended* function of the maker.

In view of the foregoing I propose the following definition of technical artefact kinds:

An object X is an instance of the technical artefact kind K iff X is the result of a largely successful execution of a largely correct design of a K.

A notable feature of this definition of technical artefact kinds is that being an instance of a technical artefact kind is determined by both the intentional and material history of the artefact involved. The intentional history refers to the intentions of the maker, in particular the function intended by the maker and the design (s)he has in mind. The material history refers to the largely successful execution of the correct substantive idea, that is, to the physical properties of the object made.

This conception of being an instance of a technical artefact kind makes it possible to deal with the malfunction and circularity problem. A technical artefact that has lost its capacity to perform the function associated with its artefact kind, that is, a malfunctioning technical artefact, may still be an instance of that technical artefact kind by virtue of its intentional and material history. Thus, a broken TV-set may still be a TV-set. Also here we encounter vagueness, not so much with regard to the substantive idea underlying the definition of a technical artefact kind, but with regard to how much physical change (damage) a once well-functioning instance of a technical artefact kind may sustain before it ceases to be an instance of that kind, that is, before the object with the relevant intentional and material history ceases to exist. It will be hard, if not impossible, to draw a sharp dividing line between changes in physical structure that do not and that do affect membership of a technical artefact kind. Leaving these issues about vagueness aside, it may be concluded that if causal-role theories of technical functions are supplemented with a theory of technical artefact kinds along the lines outlined above, they may be able to deal adequately with the malfunction problem.

Similarly, this theory of technical artefact kinds may help evolutionary theories of technical functions overcome the problem of circularity. It offers those theories a way of defining (being an instance of) a technical artefact kind that is independent of the proper functions as defined by those theories. The function that plays a role in the definition of a technical artefact kind is the function intended by the creator or maker of the artefact. This *intended* function is not to be confused with a proper function as defined by evolutionary theories of function. The circularity objection may be avoided because now we are dealing with two different kinds of functions.

4.4 Use-proper functions versus kind-proper functions

In order to keep these two functions clearly apart, I will refer to the function that is part of the substantive idea of our theory of technical artefact kinds as *kind-proper function* and the function defined by theories of functions based on use contexts as *use-proper function.* Kind-proper functions are the functions intended by the makers/creators and are assigned during processes of creation and reproduction of technical artefacts and technical artefact kinds. The kind-proper function assigned to an object during its process of creation together with its physical structure defines the technical artefact kind of which it is an instance. Use-proper functions, on the contrary, are related to use practices and are defined on the basis of those practices. I postpone a more precise definition of these two kinds of proper functions until later. Let me first explain why I think it is necessary to distinguish between these two kinds of proper functions.

The most important reason for driving a wedge between use-proper functions and kind-proper functions is that instances of a technical artefact kind may be used in (systematic) ways that deviate from the use for which these instances were originally intended and made. In such situations those instances may acquire new 'stable' functions which may become new use-proper functions that are different from their kind-proper function. These instances, however, retain their kind-proper function, because their intentional and material history 'sticks' to them. So, a new use-proper function may be acquired without a corresponding change in artefact kind.[25] Typically the use-proper and kind-proper functions will be the same; this simply reflects the fact that most technical artefacts (especially more complex technical artefacts like mobile phones, copiers, TV-sets, air planes et cetera) are standardly used in the way they are intended by their makers. Occasionally, however, the two functions may become apart. Let us first consider some typical cases.

Obsolete kind-proper functions. Consider cases where technical artefacts of some kind, that are initially being used in accordance with their kind-proper function, become obsolete because in the course of time they are outperformed by instances of competing technical artefact kinds. Take, for instance, steam engines of the Savery or Newcomen kind. These technical artefacts have lost their original use-proper function corresponding to their kind-proper function. They were designed and made for driving all kind of machinery (originally mainly pumps) and that is how they were used in the early days of steam technology. The few instances of these artefact kinds that are still left now serve another role or function. Many of them are used by museum directors to be displayed for educational or other reasons. So they are assigned a different function which may be interpreted as a new use-proper function, albeit not a technical one. However, they remain steam engines of

[25] Psychological investigations suggest that beliefs about an object being a member of an artefact kind and beliefs about the (current) function of the object do not always run parallel; see (Bloom 1996; German and Johnson 2002).

the Savery or Newcomen type, that is, they remain instances of the original technical artefact kind.[26] There are also cases in which instances of a technical artefact kind are no longer able to perform their kind-proper function due to wear and tear and are used for other purposes leading to new stable use-proper functions. Worn-out tires are a case in point; they acquire other technical use-proper function, such as the use of old tires as planters or the use of old tires for keeping in place plastic foil covering hay stacks.

Parallel use-proper functions. Kind-proper and use-proper functions may also come apart in cases in which instances of a technical artefact kind acquire a new use-proper function that exists parallel to the original use-proper function, identical to the kind-proper function, without a new artefact kind coming into existence or the artefact becoming an instance of two different technical artefact kinds. What Preston (1998, p. 251) calls *standardized ongoing exaptations*, such as the use of chairs as a stepladder, or flatirons as a doorstop or of screwdrivers as paint-can-openers may come under this heading. Mark the use of language in these cases: an instance of an artefact kind is used *as* an instance of another artefact kind, without becoming an instance of the latter kind. Of course, the definition of use-proper function has to be such that it countenances such forms of standardized use as a sufficient basis for proper function attribution. Our proposal for use-proper functions will indeed allow that.[27]

Test exemplars. Finally, there is a whole category of technical artefacts, namely test exemplars, whose members are assigned a use-proper function that is clearly different from their kind-proper function. Consider a particular kind of satellite of which only two instances are built. One, let us call it the 'communication satellite', is to be shot into space and to be used as a communication satellite once in orbit around the earth; the other, the 'test satellite', is to remain on earth to be used for testing in case something goes wrong with the one in space. Here we are dealing with two instances of one technical artefact kind with two very different use-proper functions, one of which coincides with the kind-proper function. It might be objected that the function that is part of the largely correct substantive idea on the basis of which the test satellite is made, is different from the function that is part of the largely correct substantive idea of the communication satellite and that therefore the two satellites, on my account of technical artefact kinds, are not instances of one and the same kind of technical artefact. However, making an object with the function of being a test satellite for the communication satellite implies making a truthful copy of the communication satellite. Now, the phrase 'making a truthful copy" may be interpreted as making an object that is a truthful physical copy of the material object involved in the communication satellite. On that interpretation

[26] It may be objected that the newly acquired proper function is a status function and that because of this new status function the engines are instances of the kind of museum objects. However this may be this in no way affects the kind-proper function of these engines which makes them instances of a particular technical artefact kind.

[27] Preston's reason for denying that in these cases we are dealing with proper functions is that the artefacts involved are not reproduced for this use.

making a truthful copy of the communication satellite does not involve making another instance of that technical artefact kind. Moreover, from a pragmatic point of view this interpretation of making a truthful copy is very problematic, because it leaves the makers of the physical copy without any relevance criteria for deciding when a copy is a truthful physical copy (unless, of course, they are making an exact copy, molecule for molecule, but even this notion of exact copy may be put into question on physical grounds). These relevance criteria can be provided only by taking into account the function of the communication satellite as a whole and of all of its components. This means that the makers of the test exemplar must start from the same largely correct substantive idea, including the same function, in making the test exemplar as is being used in making the communication satellite. So they are making another instance of the technical artefact kind to which the communication satellite belongs, only they have another 'proper use' in mind for the object they are making. From the point of view of the theory of technical artefact kinds presented above, there appears to be nothing problematic about the idea that a new instance of a technical artefact kind may be made with the explicit intention of using that instance in a way that is different from the kind-proper function associated with that technical artefact kind.

These examples make clear that kind-proper functions and use-proper functions are different in nature and do not always coincide. In fact, the distinction between kind-proper functions and use-proper functions corresponds to two different ways the notion of proper function is construed in the literature on function theories. According to one construal the notion is related to artefacts having functions on their own, according to the other to (standardized) use practices.

As regards the first construal, Millikan remarks that when she coined the term 'proper function' she meant the term 'proper' in the original Latin meaning of *proprium*, meaning *one's own* (1984, p. 2; 1999, p. 192). The proper function of an artefact is the function it has of its own. She stresses that nothing judgmental was meant with the term. A proper function may entail that a thing ought to have a certain capacity, but the normative term 'ought to' is not to be taken in an evaluative sense; it indicates a measure or norm to which the thing may or may not in reality conform. Griffiths (1993, p. 411) characterization of proper functions in terms of what things are *for* comes close to Millikan's when this for-ness is associated with a thing *considered on its own*. According to Griffiths proper functions are also special because they may play a role in explaining (partly) the presence of the technical artefact; this goes back to one aspect of Wright's seminal analysis of functions (1973), namely that an item X with function Z is there because it does Z.

The second construal of the notion 'proper' considers an artefact not in its own, but interprets the 'what it is for' in the context of some larger system of which it is part or of some (standardized) use practice. Especially when proper functions are coupled to (standardized) use practices, the notion of proper tends to acquire an evaluative meaning, which is reflected in expressions like "functioning properly" or "being used properly". This is, for instance, evident in Houkes and Vermaas' analysis of proper functions in terms of socially approved use plans. This explication of the notion of proper function makes the term 'proper' a heavily normative one.

The notion of kind-proper function is intended to capture the first construal of the notion of proper function. This notion is closely related to the idea that proper functions are functions of technical artefacts on their own, that is, technical artefacts considered by themselves, independently of how they are actually used. This does not mean that kind-proper functions are necessarily mind-independent properties of technical artefacts and are not assigned. Following Thomasson's account of artefacts and artefact kinds, something is an instance of an artefact kind by virtue of its intentional and material history. So considering an artefact on its own implies taking into account its intentional properties (intentional history) and the function of the artefact on its own, its proper function, may be considered to be the intended function assigned by its maker that is part of the largely correct substantive idea of that artefact kind. This proper function is inherent to the artefact *qua instance of its artefact kind*; by definition this proper function, together with some core design features, makes the object into an instance of that kind. Moreover, this proper function may play a role in explaining the presence of a particular artefact, as an instance of an artefact kind; its presence is due to the successful execution of a largely correct substantive idea of its artefact kind, which includes to idea of the intended function, that is, the kind-proper function.[28]

The distinction between use-proper and kind-proper functions implies that some use-proper functions are 'more proper' than others, because they coincide with kind-proper functions.[29] The use-proper function that coincides with the kind-proper function is special for two different reasons. The first is that, in line with Griffiths' remark, this use-proper function is the *raison d'être* of the artefact kind in the sense that it's first token was designed and produced to be used for that function. That does not exclude other (standardized) ways of using it. However, it is doubtful that instances of a certain kind of artefact would still be manufactured and would remain available on the market the moment its kind-proper function would become obsolete. As Preston remarks with regard to the use of chairs as a stepladder (1998, p. 241):

> So if chairs were no longer manufactured for their proper function of supporting seated humans, the most plausible consequence is that they would just disappear from the artifactual repertoire of our culture, while step stools and stepladders would proliferate.

This explains why typically the use-proper function of instances of an artefact kind corresponds to the kind-proper function.

The second reason why the use-proper function identical to the kind-proper function is special is not at all related to actual use practices, but to the way kind-proper

[28] Although she does not refer to artefact kinds but to function categories, also Millikan (1993, p. 21–22) connects the issue of proper functions to being a member of a function category (artefact kind). Members of a function category are not always able to perform the proper function associated with that category. One of the problems that a definition of proper functions is supposed to solve is (*ibidem*) "how did the atypical members of the category that cannot perform its defining function *get* into the same function category as the things that actually can perform the functions?".

[29] See also (Ridder and Kroes 2006).

functions are defined and to one of the core values of engineering practice. According to Thomasson somebody creates an instance of an artefact kind when (s)he more or less successfully executes substantive intentions to create something of that kind. For this it is necessary that the maker has a largely correct substantive idea of what an artefact of that kind is. In line with the dual-nature conception of technical artefacts I have argued that for technical artefacts this means that the maker must have a largely correct idea of its core functional and design features, which amounts to having a largely correct idea of its design.

Minimally, the notion of a largely correct design implies that an artefact built according to that design will be effective, that is, realizes the intended or kind-proper function involved. But from an engineering point of view much more is involved in being a largely correct design than just effectiveness. Apart from effectiveness, efficiency plays a dominant role. Within engineering there is a strong drive towards efficiency, that is, to realize functions (goals) with a minimum of resources (see Alexander (2009)). So a largely correct design will in general also have to satisfy conditions of efficiency. It is precisely from the perspective of this drive towards efficiency that the use-proper function corresponding to the kind-proper function may be expected to be in a special position. Since an artefact of a certain kind is designed and manufactured for use in accordance with its kind-proper function, it may be expected that, when it is designed in an effective and efficient way, it does not contain many parts and features that are irrelevant for realizing its function. With regard to kind-proper functions something like a Principle of Maximal Component Utilization (MCU) seems to be at work (Ridder 2007, p. 214): almost all of an artefact's components and many of their specific features contribute directly or indirectly to some aspect of the artefact's functioning, use, maintenance, or assembly. For use-proper functions that are different from the kind-proper function, there is no reason to assume that this principle applies. Surely the principle does not apply when it comes to the structure of tires and their use as planters or for keeping plastic foil in place on haystacks. So also from the perspective of how the structure and function of a technical artefact are related to each other, the use-proper function corresponding to the kind-proper function appears to have a special place among all possible use-proper functions.

Finally, let me return to some of the problems I discussed with regard to Preston's and Houkes and Vermaas' theories of proper functions and indicate briefly how the distinction between use- and kind-proper function and our concept of technical artefact kinds may help overcome them.[30] With regard to Preston's theory I have already pointed out that the distinction between use- and kind-proper function opens the possibility to avoid the charge of circularity: being an instance of an artefact kind can be defined independently of the use-proper function that is defined in terms of use and reproduction histories. The notion of kind-proper function also makes it possible to attribute kind-proper functions to prototypes and one-of-a-kind technical artefacts, although, on Preston's account, it would not be possible to attribute to

[30] I disregard Searle's theory since it does not address the issue of proper functions explicitly.

them use-proper functions, because of a lack of use and reproduction history. As far as Houkes and Vermaas' theory is concerned, the adoption of the notion of kind-proper function makes it likewise possible to effectively deal with problems about proper functions of prototypes and one-of-a-kind technical artefacts. More importantly, it allows the theory to deal with cases of "known" malfunctioning. Consider again the TV-set that is broken and of which the user knows that it is broken. On the ICE-theory it is not possible to ascribe the function of producing TV-images to this TV-set. Given the assumption that the ascription of this function is necessary to be an instance of the artefact kind TV-set, that leads to the conclusion that that object is not a TV-set. On our account of technical artefact kinds and use and kind-proper functions, however, that object is still a TV-set because it has the relevant kind-proper function, and normative statements about the TV-set can be based on this kind-proper function. The broken TV-set, however, would have no use-proper function, if the latter is defined the way the ICE-theory defines proper functions.

4.5 Three kinds of technical functions

Taking into account the distinction between use-proper functions and kind-proper functions, at least the following three kinds of functions may be distinguished with regard to technical artefacts.

Use-accidental functions. Use-accidental functions are the weakest or most unstable kind of technical functions. They are assigned to technical artefacts or natural objects in relation to accidental use, that is, use that is not intended or standardized by way of use plans sanctioned within a community of users. This use differs from use according to its use-proper function(s) or kind-proper function (in case we are dealing with technical artefacts). These functions may be analysed in terms of Cummins-style causal-role/system functions: a technical artefact *X* is part of a system *S* and contributes to an overall capacity of *S* by its capacity to φ. Take, for instance, the system consisting of an agent, a closed paint can and a large nail; assuming that the agent has appropriate skills the overall system has the capacity to open this paint can by using the nail as a tin can opener. As soon as *X* is taken out of the system, the accidental function disappears; nothing of this function 'sticks' so to speak to the object *X* in isolation from the system *S*; considered on its own, the nail does not have the function of opening tin cans, nor is it a member of the kind of tin can openers. Because of this it is generally assumed that the assignment of use-accidental functions cannot play a role of any significance in ontological theories of functions or technical artefact kinds. Accidental use of a nail as a tin can opener does not turn that object into an instance of the technical artefact kind of tin can opener.

Use-proper functions. Use-proper functions are more stable than use-accidental functions. Technical artefacts may acquire use-proper functions in various ways, which may lead one to introduce different notions of use-proper functions. One obvious way is by more or less standardized use practices (use plans). In this case use-proper functions are assigned to or acquired by technical artefacts or natural objects (think of the example of shells used as drinking cups) in relation to a community of users in which they are used in a more or less standardized way. For technical artefacts these use-proper functions as a rule coincide with their kind-proper functions, that is, the function that is part of the largely correct substantive idea of their designers and makers. But these use-proper functions may also be different from kind-proper functions (and may therefore be taken to be accidental from the point of view of the kind-proper function). In contrast to use-accidental functions, a *community* of users is now involved in the assignment of use-proper functions; formal or informal social rules or institutions may lie at the basis of use-proper functions (Scheele 2006). No individual user is able, by herself, to introduce a new use-proper function; social elements play a crucial role in this kind of proper functions.[31] Technical artefacts may also acquire use-proper functions when an instance of a technical artefact kind is made with the explicit purpose to be used in a way that is different from its kind-proper function. Think of the example of the test satellite. Its use-proper function, the function it is intended for, is to be used for testing not communication. This kind of use-proper function is not necessarily related to a community of users but presupposes a specific context of making and use.

Use-proper functions are not proper in the sense that they are functions of technical artefacts considered on their own. They are defined relative to a community of users or to a specific context of making and use. If we take them out of this community of users, or out of this specific context of production and use, these technical artefacts loose their use-proper functions (in case the use-proper functions are identical to the kind-proper functions, this does not imply that they also loose their kind-proper function). This may be interpreted as implying that the assignment of use-proper functions has no ontological significance, because they do not result in adding new technical artefacts to the world. Finally, note that for use-proper functions defined relative to a community of users there is no sharp dividing line between use-accidental and use-proper functions; the distinction is one of degree and depends on the level of standardization of use within a community of users.

Kind-proper functions. Kind-proper functions are assigned by agents in the course of making a first instance of a new artefact kind or in the course of making more instances of an already existing artefact kind. They are part of the largely correct substantive idea of a technical artefact kind that is involved in these processes. I take the assignment of kind-proper functions to have ontological significance because it leads to new kinds of technical artefacts or to new instances of already existing kinds of artefacts. Kind-proper functions are the only kind of functions that are

[31] A direct consequence of our characterization of the various kinds of functions is that the individual use of an artefact in accordance with its kind-proper function may have to be interpreted, paradoxically, as the assignment of a use-accidental function. This may happen when the actual use-proper function is different from the kind-proper function. In a community in which the use of flatirons for ironing has become obsolete and in which their use as door stops is standardized, an individual using a flat iron for ironing shirts assigns by that use a use-accidental function to the flat iron.

properly called 'proper' in case that term is taken to mean 'of its own'. Kind-proper functions stick to technical artefacts independently of their context of use.[32] The way kind-proper functions are defined implies that whether an artefact is an artefact of a particular kind is determined by its intentional and material history, in particular its design and production history.

These three kinds of functions have different normative features. In the function literature normativity is usually associated with proper functions only. It is generally assumed that accidental or system functions cannot be the basis for 'ought to' statements and that objects with these kinds of functions cannot malfunction, since they lack the function when they do not have the relevant capacity. In my opinion normative statements may be made in connection with all three kinds of functions, but they vary greatly in their normative strength and these normative statements do not always correspond to malfunctioning statements.

Let us start with use-accidental functions. Suppose that I know that the on/off switch of my TV-set is out of order and I decide to use a screwdriver to make contact between the two wires of the switch in order to 'switch' it on. By so using it I assign the screwdriver the systemic function of on/off switch; I use it as an on/off switch. I clean the shaft of the screwdriver, notice that it is not covered with insulating plastic, is made of metal (a good conductor for electric current) and position (the tip of) the shaft such that it touches the two wires connected to the on/off switch. Irrespective of whether this trick works or not, it makes perfect sense for me to say "This ought to work" as a kind of shorthand for "This screwdriver ought to work as an on/off switch". In case the result is negative it does not make sense to call the screwdriver a malfunctioning on/off switch (even to say that it malfunctions as an on/off switch sounds rather odd). Nevertheless we may associate 'ought to' statements with use-accidental functions. In line with the conclusion we reached earlier, however, there is nothing intrinsically normative about this kind of functions, since such 'ought to' statements may simply be interpreted in terms of reasonable, but defeatable expectations about the behaviour of the screwdriver in that specific context (system).

When we move on from use-accidental to use-proper functions, the situation with regard to normativity does not change very much. It is obvious that in case of use-proper functions based on more or less standardized use practices, the difference between a use-proper and use-accidental function is a matter of degree, depending on the degree of standardization of a particular use in a community of users. Changing the scene from an individual agent to a community of users, however, changes the kind and strength of evidence for the expectations which are

[32] With the help of the distinction between use-proper function and kind-proper function we may explain why instances of a technical artefact kind, of which almost all instances are and will never be used (for instance, because they are outperformed by a competitor the moment they were produced) may still have a proper function, namely a kind-proper function, and thus be an instance of the technical artefact kind. This is a problem similar to the one in biology where only a tiny fraction of an item type may actually perform its proper function (e.g., sperm cells); see for instance (Neander 1991).

expressed in the form of 'ought to' statements. Social epistemology enters the stage, which means that expectations about the behaviour of artefacts may be based not only on personal experience (as in the above example) but also on the experience and testimony of others.[33] New kinds of evidence become available; so, the evidence base for 'ought to' statements enlarges which means that the 'ought to' statements based on this evidence base may gain in strength. Moreover, some of the 'ought to' statements may have a moral nature, instead of an epistemological one. Strictly speaking, only in case the use-proper function of an artefact coincides with its kind-proper function, does it make sense to speak of a malfunctioning artefact when the use of the artefact fails to bring about the desired effect (an old tire whose use-proper function is to keep the plastic foil on a hay stack in place but fails is not a malfunctioning tire).[34]

Only with regard to kind-proper functions malfunctioning may occur: an instance of a technical artefact kind that is unable to perform its kind-proper function is a malfunctioning instance of that kind. Independently of its actual use, that is, of the use-accidental or use-proper function actually assigned to it, a technical artefact considered on its own ought to be able to perform its kind-proper function. This is a kind of 'ought to' that appears to be inherent to the artefact involved. This makes it a strong kind of normativity because it is independent of any use context. This kind of normativity may be inherent in the sense that it belongs to an artefact considered on its own, but not intrinsic in the sense of mind-independent, since being an artefact of a certain kind is a mind-dependent property. The two success criteria in the conception of what a technical artefact of a certain kind is, imply that when a person is confronted with an instance of a technical artefact kind she is justified in or has reasons for expecting that this instance will perform its kind-proper function.[35] Of course, expectations may be defeated because a technical artefact has some defect. However, as long as that defect does not undermine the statement that the technical artefact is an instance of a certain technical artefact kind, the kind-proper function of the artefact has normative implications, which may be interpreted in terms of justified expectations about its behaviour. So, in line with our conclusion about the nature of the normativity of functions, also the inherent normativity associated with kind-proper functions may be taken to be grounded in the normativity of human agency.

These three kinds of technical functions play a crucial role in the way I intend to combine a theory of technical functions with a theory of technical artefact kinds. Before I turn to an outline of such a combined theory, I summarize the main results obtained so far and draw up a list of adequacy conditions that a combined theory will have to satisfy.

---

[33] See also Houkes and Vermaas' distinction between professional and amateur designing (2010, p. 27).

[34] An analogous case in biology are the 'flippers' of penguins; they are not malfunctioning wings.

[35] For a discussion of the various success criteria involved in evaluating artefacts, see also Hilpinen (2004, 1992).

4.6 Adequacy conditions

The main outcomes of our analysis of the nature of technical functions and of technical artefact kinds upon which our proposal for a combined theory is based are the following ones. The first is that functions of technical artefacts are mind-dependent properties; technical artefacts have functional properties only in relation to intentional human action. Second, this mind-dependence finds its origin in performative function ascriptions (that is, function assignments) that have to be carefully distinguished from descriptive function ascriptions). Third, technical artefacts perform their function on the basis of their (physical/material) structure. That is what makes technical artefacts different from social artefacts; the latter perform their functions on the basis of collective intentionality. These three results together lie at the bottom of the dual-nature conception of technical artefacts. A fourth outcome is that there is nothing intrinsically normative about technical artefacts or technical functions. In so far normative statements can be made with regard to technical artefacts, this normativity has to be interpreted in terms of the normativity inherent in human agency. Finally, the use-proper function of a technical artefact may be different from its kind-proper function. The latter finds its origin in the intentions of its creator or maker.

The following list of adequacy conditions for theories of technical functions and technical artefact kinds is inspired by the list of desiderata proposed by Vermaas and Houkes for theories of technical functions (see section 3.5.3). Apart from adapting their desiderata I also add two more, one about the distinction between functional and causal effects and one about the relation between proper functions and being an instance of a technical artefact kind. Here is my list.

> *A1. The functional versus causal-effect condition*: A theory of technical functions and technical artefact kinds should distinguish functional effects from mere causal effects.[36]

I add this condition to the list of Vermaas and Houkes because causal-role/systemic theories of function have been charged with the 'promiscuity' objection. Almost any causal effect of a technical artefact may become its function by embedding it in a suitable encompassing system (for instance, the monitor in front of me is part of the system of my office that has the function of offering me a good working space; one of the causal effects of the monitor is to produce heat and therefore its function may be taken to heat my office; another causal effect of the way the monitor is positioned is that it blocks part of the outside garden from my view; its function may therefore be taken to prevent me from getting distracted from my work by what is going on in the garden; but the monitor is also part of the system

[36] See Davies (2001, p. 103).

consisting of all 17-inch flat monitors on earth and its function is to contribute to the overall weight of this system etcetera).[37] This leads to an undesirable proliferation of functions of technical artefacts; without any constraints on the kind of allowable encompassing systems, functional and causal effects may become, it seems, more or less co-extensive.

A2. The proper versus accidental function condition: A theory of technical functions and technical artefact kinds should make it possible to distinguish between accidental and proper functions of a technical artefact.

Among functions (functional effects), it is common to distinguish between accidental and proper functions. If someone on purpose positions my monitor such that it blocks my view of part of the garden, then the function of blocking my view is an accidental and not the proper function of this monitor. The function assigned to the monitor is considered to be accidental because that is not what monitors are for. The validity of warranty conditions on technical artefacts usually depends on proper use of the artefact, that is, roughly, use in accordance with its proper function. Some aspects of the distinction between proper and accidental functions are related to the notion of technical artefact kinds, since the proper function of a technical artefact is commonly taken to be the function that is "essentially associated" with the technical artefact kind to which it belongs (Von Wright 1963, p. 20).

A3: The malfunction condition: A theory of technical functions and technical artefact kinds should admit the possibility of malfunctioning technical artefacts, that is, should admit the possibility that an artefact is an instance of a technical artefact kind, without that object being able to perform the proper function associated with that technical artefact kind.

This malfunction condition is intended to cover normative aspects of technical functions in general: a theory of technical functions and technical artefact kinds should be able to account for the possibility of normative/evaluative statements about technical artefacts.

A4: The function-structure coherence condition: A theory of technical functions and technical artefact kinds has to take into account that the structure and function of technical artefacts constrain each other in the sense that, generally speaking, it will not be possible to change the function of a technical artefact without changing its structure, and vice versa.

[37] For a discussion of the promiscuity objection to systemic function theories, see Davies (2001, Ch. 4).

This condition states that there is a coherence between structure and function of technical artefacts (Kroes 2006). Although the wording is different, this condition expresses the same intuition that lies at the bottom of Vermaas and Houkes' physical structure condition, namely that not every artefact can perform a given function. For characterizing technical functions and technical artefact kinds this condition is rather crucial since it is this feature that distinguishes technical artefacts from social artefacts.

A5: The novelty condition: A theory of technical functions and technical artefact kinds should allow that innovative technical artefacts (first instances of new artefact kinds) and one-of-a-kind technical artefacts have proper functions (whether 'have' is taken in an epistemological or ontological sense).

Theories that do not satisfy this condition are not in accordance with functional discourse in general, and with engineering practice in particular.

A6: The change in proper function condition: A theory of technical functions and technical artefact kinds should allow that instances of a technical artefact kind may change their proper function without a change in their artefact kind.

In discussions of theories of technical functions it is often pointed out that technical artefacts may systematically be used in ways not intended by their designers and/or makers and that this may lead to a change in their proper functions, either at the level of instances of the kinds (different proper functions in different phases of their life cycle) or at the level of the artefact kind (all instances of the artefact kind change their proper function by being used for a different purpose). A combined theory should be able to account for these phenomena.

4.7 A theory of technical functions and of technical artefact kinds

Finally, we are in a position to propose epistemic and ontological theories that explicate what it means for objects to 'have' the functional properties of being for φ-ing and being a φ-er, that is, being an instance of the technical artefact kind φ-er. The combination of theories of technical functions and technical artefact kinds that follows consists of a set of explicative definitions that introduce and fix the meaning of various notions that in my opinion are necessary to account for, and where appropriate refine and revise, our discourse about technical functions, technical artefacts and technical artefact kinds. In other words, these combined theories are intended to offer a conceptual framework for interpreting statements about objects having functional properties. This framework will have to satisfy the adequacy conditions formulated in the previous section. Moreover, this conceptual framework is intended to make it possible to account for and further explicate the dual nature of technical artefacts.

Let me start with presenting the ontological theory of object x being an instance of a particular technical artefact kind and x having the corresponding kind-proper function:

1a) Object x has the functional property of being a φ-er (or: is an instance of the technical artefact kind φ-er) iff x is the result of a largely successful execution of a largely correct design of a φ-er.
1b) Object x has the kind-proper function of being for φ-ing if it has the functional property of being a φ-er.

Several points are to be noted. First, in section 3.2 I assumed that the functional property of being a φ-er implies having the functional property of being for φ-ing, but not the reverse, since an object may perform an accidental function without becoming an instance of the corresponding technical kind. The distinction between kind-proper functions and use functions makes it possible to be more precise on this point. Having the functional property of being a φ-er implies having the kind-proper function of being for φ-ing (1b). And indeed, having the use-proper or use-accidental function of being for φ-ing does not imply the functional property of being a φ-er. The same applies also to having the kind-proper function of being for φ-ing; that does not imply that the object involved is an instance of the technical artefact kind φ-er. However, it does imply that the object is an instance of the technical kind consisting of all technical artefact kinds that have the kind-proper function of being for φ-ing in common. The reason for this is that the kind-proper function of being for φ-ing is multiple realizable, and so having that property does not imply being an instance of the technical artefact kind φ-er, since the latter artefact kind is not only determined by its kind-proper function but also by the particular physical realization fixed in its design (1a).

Second, this definition does not yet uncover the ontological role of function assignments. This role is hidden in 1a. I assume that the successful execution of a largely correct design of a φ-er involves an ontologically significant assignment of the kind-proper function being for φ-ing by the maker. Thus, somebody who meticulously copies the physical structure of a given technical artefact, without any understanding of its function or the functions of its components, does not create another instance of that technical artefact kind. Because of this pivotal role of function assignments in making an instance of a technical artefact kind, technical artefacts are ontologically mind-dependent objects.

This brings me to a third point that concerns the success criteria for creating an instance of a technical artefact kind. According to the above definition the largely successful execution of a largely correct design is a necessary and sufficient condition. This implies that a technical artefact made on the basis of a principally flawed design (an incorrect substantive idea) may be an artefact, but is not a *technical* artefact. Similarly, an object produced on the basis of a largely correct design of a technical artefact kind, but executed in a lousy way, is not an instance of that technical artefact kind. From this success criterion the criteria have to be derived that are to be imposed on function assignments in order to warrant their ontological effect. As I suggested in section 3.5.3, the ICE-theory of Houkes and Vermaas for function ascriptions by designers, when function ascriptions are interpreted in a performative sense as function assignments, may be taken as an attempt to spell out the conditions for ontologically significant function assignments.[38]

The epistemic counterpart to the above ontological theory states the conditions for justified technical artefact kind ascriptions (being a φ-er) and of the corresponding kind-proper function ascription (being for φ-ing):

2a) An agent *a* justifiably ascribes to *x* the property of being a φ-er (being an instance of the technical artefact kind φ-er) iff *a* justifiably believes that *x* is the result of a largely successful execution of a largely correct design of a φ-er.

2b) An agent *a* justifiably ascribes to artefact *x* its capacity to φ as its kind-proper function if *a* justifiably believes that *x* is a φ-er.

From an epistemic point of view different situations may have to be distinguished. Kind or kind-proper function ascriptions may be done by (i) the creator of the first instance of a new technical artefact kind, or (ii) by someone who intends to make another instance of an already existing technical artefact kind, and (iii) an observer. As Thomasson has argued the creator of a first instance of a new artefact kind may be in an epistemically privileged position. Since a successful execution of a largely correct design involves a function assignment, ascriptions of the functional properties of being a φ-er or having the kind-proper function being for φ-ing are based on beliefs about kind-proper function assignments that are assumed to have ontological significance. This makes this epistemic theory, as was to be expected, a mind-dependent one.

For the definition of the ascription of use-proper functions I will bring into play a slightly modified version of Hansson's tentative definition of a "socially recognized technical function" (Hansson 2006b, p. 22). Hansson's definition is largely based on a recombination of various elements contained in the ICE-theory that focuses on individual function ascriptions.

3) An agent or group of agents justifiably ascribes to an artefact *x* the *use-proper function* of being for φ-ing relative to a use plan *p* for *x* iff:

(1) *p* has been assigned to *x* by an agent or a group of agents that has the appropriate type of social position in relation to *x*, and
(2) it is both true and socially recognized that there is considerable chance that competent execution of *p* with *x* will lead to the goal of φ-ing.

---

[38] Note that in that case the capacity and contribution beliefs of the designers have to be true and not only justified on the basis of some account A. This is necessary to exclude the possibility of an incorrect (principally flawed) yet justified design.

As Hansson remarks, such a definition leaves a number of points to be further developed. As always questions about borderline cases may be raised, for instance concerning the minimum size of the social group or community involved. Houkes and Vermaas (2010) address issues about what an "appropriate type of social position" might be by distinguishing between various agent roles, among which the roles of designer and justifier. Note that a use-proper function *ascription* is based on the (social) *assignment* of the use-proper function if we assume that assigning a use plan to an object includes a function assignment.

If we remove in an appropriate way the social aspects from the definition of the ascription of use-proper functions we end up with the definition of the ascription of use-accidental functions.

4) An agent justifiably ascribes to an artefact x the *use-accidental function* of being for φ-ing relative to a use plan p for x iff:

   (1) p has been assigned to x by the agent, and
   (2) it is both true and recognized by the agent that there is considerable chance that competent execution of p with x will lead to the goal of φ-ing.

Note that, just as the use-proper function ascribed to a technical artefact may be (and usually is) identical to its kind-proper function, this may be the case for the use-accidental function ascribed to a technical artefact. This is not as strange as it may seem: an agent, without knowing what kind of technical artefact he is dealing with, may by chance ascribe it a use-accidental function that is identical to its kind-proper function.

For reasons discussed earlier, I assume that there are no ontological counterparts to the ascriptions of use-proper and use-accidental functions. I take the function assignments on which these function ascriptions are based to have no ontological significance. The technical artefacts involved do not undergo any change in artefact kind. The assignment of the function of driving screws to a coin by an individual agent or a group of agents, how successful it may be, has no ontological impact and cannot form the basis for an epistemically objective ascription of the kind-proper function of driving screws to the coin (that is, for claiming that the coin is a screwdriver). The basic idea underlying this point of view is that generally speaking using an object does not enrich the ontological structure of the world, whereas designing and making technical artefacts does. It does not matter much whether the object involved is a natural object or a technical artefact that is used in a way that deviates from its kind-proper function. In both cases, using an object means appropriating an object that already exists for a purpose one has in mind. Historically it may be the case that the roots of making technical artefacts may be traced back in a continuous way to using (appropriating) natural objects without any or only minor modifications for realizing practical ends. From such a perspective, making technical artefacts may emerge as the (almost purely) intellectual activity of successfully projecting functions on objects at hand. Such a view of making new technical artefacts kinds, however, appears hardly applicable to how new technical artefacts come into being in modern technology, since the objects with the relevant physical capacities are not lying around to be selected and appropriated for technical purposes. The largely successful execution of the largely correct designs involved in bringing about new technical artefact kinds involves much more, as we will see in the next chapter, than simply making a clever choice of what is already at hand.

The combination of the theory of technical functions and technical artefact kinds allows me to resolve the puzzle about the role of the successful performance of a function in function ascriptions (see section 2.5.2). There I concluded that, on the assumption that justified function ascriptions determine being an instance of a technical kind, successful performance cannot be a necessary nor a sufficient condition for justified function ascriptions. It could not be necessary because of malfunctioning technical artefacts, and not sufficient because of successful accidental use. So, what role, if any, does the successful performance of functions play in justified functions ascriptions? From an engineering point of view it is rather obvious that this successful performance has to play a major role. To solve this puzzle, it has to be realized that the assumption that function ascriptions determine technical kinds is ambiguous, since various kinds of function may be involved in a function ascription. From definitions (3) and (4) it is immediately apparent that successful performance plays a crucial role in the ascription of justified use-proper and use-accidental functions. For the ascription of a use-accidental function successful performance even appears to be a sufficient condition. Thus, when a coin is successfully used as a screwdriver, the ascription of the accidental function to drive screws is justified. However, this does not imply that the coin is an instance of the technical kind screwdriver. The assumption that function ascriptions determine technical kinds does not hold for the ascription of use-accidental functions. It only holds for ascriptions of kind-proper functions. Thus, successful performance of a function may be a sufficient condition for the ascription of a use-accidental function. So, the argument about the successful use of the coin as a screwdriver in section 2.5.2 is seriously flawed. The same is true for the argument that successful performance cannot be a necessary condition for the ascription of functions because of malfunctioning technical artefacts. Indeed, an object may be ascribed a kind-proper function even when it is not able to successfully perform that function. It is then a malfunctioning instance of a technical artefact kind. However, (1) implies that for making an instance of a technical artefact kind, the largely successful performance of the corresponding kind-proper function by the object made is a necessary condition. In the course of its life, that technical artefact may become damaged such that it is no longer able to perform its kind-proper function successfully. Nevertheless, according to (2) it may be justifiably ascribed its kind-proper function, and therefore be ascribed the property of being an instance of the corresponding technical kind, since (2) refers to the intentional and material history of the technical artefact, not to its actual physical properties.

Let me now briefly confront the combined theory with our list of adequacy conditions. The first condition (A1) requires that it is possible to distinguish between causal and functional effects. Only causal effects that play a role in some plausible function assignment may figure as a functional effect. In (4), the weakest form of function ascription, the unrestrained proliferation of functional effects is blocked by reference to the notion of use plan as proposed by Houkes and Vermaas; this reference restricts functional effects to what may be called reasonable use plans. The proper versus accidental function condition (A2) is taken care of by the distinction between kind-proper functions on the one hand and use-proper functions and use-accidental functions on the other. This distinction not only reveals an ambiguity in the notion of proper function used in formulating the proper-accidental condition but it also shows that the traditional opposition between proper and accidental functions may be construed in different ways. Our interpretation of technical artefact kinds makes it possible to account for the fact that a technical artefact that is unable to perform its kind-proper function may still be an instance of its technical artefact kind (A3). The function-structure coherence condition (A4) is taken care of by the requirement of the successful execution of a largely correct design and by the conditions imposed on use-proper and use-accidental function ascriptions in the second clause of respectively (3) and (4). Our conception of technical artefact kinds and kind-proper functions makes it possible to meet the novelty condition (A5) as we have already pointed out several times. Finally, the distinction between kind- and use-proper function allows accounting for changes in (use) proper functions without changes in technical artefact kinds (A6).

4.8 Conclusion

Let me recapitulate my efforts to understand the nature of technical artefacts. I have argued that in order to describe technical artefacts we have to make use of two conceptual frameworks, one for describing physical phenomena and one for describing intentional (social) phenomena. The reason for this is that the function of a technical artefact is on the one hand closely related to its physical structure, on the other hand to human intentions. On this basis I concluded that technical artefacts have a dual nature and that in order to do justice to and explicate this dual nature we need a hybrid theory of technical functions in which both physical features and human intentions play a role. I have analysed the general form of epistemic and ontological theories of two functional properties, namely of being for φ-ing and being a φ-er. Following Hansson I have pointed out the distinction between descriptive and performative senses of function ascriptions. Performative function ascriptions, called 'function assignments', play a crucial role in mind-dependent theories of functions because they connect the functions of technical artefacts to human intentions. An analysis of existing theories of technical function showed that some of the major problems they face (the malfunction problem, the circularity problem, the proper functions of prototypes) are related to the, often implicit, assumption that the notion of technical functions defined by those theories determines technical kinds. This led me to the conclusion that what is needed is not only a (hybrid) theory of technical functions but also a theory of technical artefact kinds. As my starting point I have chosen Thomasson's theory of artefact kinds and amended it for technical artefacts. One of the interesting features of Thomasson's theory is that it allows the definition of the proper function of a technical artefact in terms of the intentions of its maker. Most theories of technical functions, by contrast, define proper functions from a user or reproduction perspective. I have argued that proper functions defined in user- or reproduction-centred theories of functions have to be distinguished from proper functions defined in theories of technical artefact kinds and that the main problems faced by existing theories of technical functions may be resolved by taking this distinction into account. Finally, I have presented a combined theory of technical functions and technical artefact kinds.

The overall picture of technical artefacts that emerges from the theories proposed is that they have a dual nature: a technical artefact is constituted on the one hand by a function assignment (human intentions) during a successful execution of a largely correct design of a technical artefact, and by a concrete physical structure that is a largely successful embodiment of that design. On this picture technical artefacts are indeed creations of mind and matter.

References

Alexander, Jennifer K. 2009. The concept of efficiency: an historical analysis. In *Philosophy of technology and engineering sciences*, ed. Anthonie Meijers, 1007–1030. Amsterdam: Elsevier.

Baker, Lynne Rudder. 2008. The shrinking difference between artifacts and natural objects. *American philosophical association newsletter on philosophy and computers* 7 (2):2–5.

Bloom, Paul. 1996. Intention, history, and artifact concepts. *Cognition* 60 (1):1–29.

Dancy, Jonathan. 2000. *Practical reality*. Oxford: Oxford University Press.

Dancy, Jonathan. 2006. The thing to use. *Studies in History and Philosophy of Science* 37:58–61.

Davies, Paul Sheldon. 2001. *Norms of Nature*. Cambridge, MA: MIT Press.

Elder, Crawford L. 2007. On the place of artifacts in ontology. In *Creations of the mind: theories of artifacts and their representations*, eds. Eric Margolis, and Stephen Laurence, 33–51. Oxford: Oxford University Press.

Enç, Berent. 1979. Function attributions and functional explanations. *Philosophy of Science* 46: 343–365.

Franssen, Maarten. 2006. The normativity of artefacts. *Studies in History and Philosophy of Science* 37 (1):42–57.

Franssen, Maarten. 2009a. Artefacts and normativity. In *Handbook of the philosophy of technology and the engineering sciences*, ed. Anthonie Meijers. Elsevier Science.

Franssen, Maarten. 2009b. The inherent normativity of functions in biology and technology. In *Functions in biological and artificial worlds: comparative philosophical perspectives*, eds. Ulrich Krohs, and Peter Kroes, 103–125. Cambridge Mass.: MIT.

Franssen, Maarten. forthcoming. The goodness and kindhood of artefacts. In *Norms and the artificial; moral and non-moral norms in technology*, eds. M.J. de Vries, S.O. Hansson, and Anthonie Meijers. Springer.

German, Tim P., and Susan C. Johnson. 2002. Functions and the origins of the design stance. *Journal of cognition and development* 3 (3):279–300.

Griffiths, Paul E. 1993. Functional analysis and proper functions. *British Journal for the Philosophy of Science* 44:409–422.

Hansson, Sven Ove. 2006a. Category-specified value statements. *Synthese* 148:425–432.

Hansson, Sven Ove. 2006b. Defining technical function. *Studies in History and Philosophy of Science* 37:19–22.

Hilpinen, Risto. 1992. On artifacts and works of art. *Theoria* 7:58–82.

Hilpinen, Risto. 2004. Artifact. In *The Stanford Encyclopedia of Philosophy (Fall 2004 Edition)*, ed. Edward N. Zalta. (editor): http://plato.stanford.edu/archives/fall2004/entries/artifact/.

Houkes, W., and A. Meijers. 2006. The ontology of artefacts: the hard problem. *Studies in History and Philosophy of Science* 37 (1):118–131.

Houkes, Wybo, and Pieter E. Vermaas. 2010. *Technical functions: on the use and design of artefacts*. Philosophy of engineering and technology. Dordrecht: Springer.

Kroes, P. 2003. Screwdriver philosophy; Searle's analysis of technical functions. *Techné* 6 (3):22–35.

Kroes, P. 2009. Technical artifacts, engineering practice, and emergence. In *Functions in Biological and Artificial Worlds; Comparative Philosophical Perspectives*, eds. Ulrich Krohs, and Peter Kroes, 277–292. Cambridge, Mass.: MIT Press.

Kroes, Peter. 2006. Coherence of structural and functional descriptions of technical artefacts. *Studies in History and Philosophy of Science* 37 (1):137–151.

Kroes, Peter, and Pieter Vermaas. 2008. Interesting differences between artifacts and natural objects. In *APA newsletter on philosophy and computers*.

McLaughlin, Peter. 2009. Functions and norms. In *Functions in biological and artificial worlds: comparative philosophical perspectives*, eds. Ulrich Krohs, and Peter Kroes, 93–102. Cambridge Mass.: MIT.

Meijers, A. W. M.. 2001. The relational ontology of technical artefacts. In *The empirical turn in the philosophy of technology Research in philosophy and technology, Vol 20 (series editor Carl Mitcham)*, eds. P. A. Kroes, and Meijers A. W. M., 81–96. Amsterdam: JAI/Elsevier.

Millikan, Ruth. 1984. *Language, Thought and Other Biological Categories*. Cambridge, MA: MIT Press.

Millikan, Ruth. 1993. *White Queen Psychology and Other Essays for Alice*. Cambridge, MA: MIT Press.

Millikan, Ruth. 1999. Wings, spoons, pills and quills: a pluralist theory of function. *Journal of Philosophy* 96 (4):192–206.

Neander, Karen. 1991. Functions as selected effects: the conceptual analyst's defense. *Philosophy of Science* 58 (2):168–184.

Preston, Beth. 1998. Why is a wing like a spoon? A pluralist theory of function. *Journal of Philosophy* 95 (5):215–254.

Raz, Joseph. 1990 (1975). *Practical reason and norms*. Princeton: Princeton University Press.

Ridder, Jeroen de. 2007. Reconstructing design, explaining artefacts; philosophical reflections on the design and explanation of technical artefacts. Delft University of Technology, Delft, The Netherlands.

Ridder, Jeroen de, and P. Kroes. 2006. Why some proper functions are more proper than others. Manuscript.

Scheele, Marcel. 2006. Social aspects of technical functions: Proper function and proper use. *Studies in History and Philosophy of Science* 37 (1):23–36.

Searle, John. 1995. *The Construction of Social Reality*. London: Penguin Books.

Simons, Peter. 1995. Artefact. In *Companion to metaphysics*, eds. J. Kim, and E Sosa, 33. Oxford: Blackwell.

Soavi, Marzia. 2009. Realism and artifact kinds. In *Functions in biological and artificial worlds; comparative philosophical perspectives*, eds. Ulrich Krohs, and Peter Kroes, 185–202. Cambridge, Massachusetts: The MIT Press.

Thomasson, Amie L. 2003. Realism and human kinds. *Philosophy and Phenomenological Research* 67 (3):580–609.

Thomasson, Amie L. 2007. Artifacts and human concepts. In *Creations of the mind: essays on artifacts and their representations*, eds. Stephen Laurence, and Eric Margolis, 52–73. Oxford: Oxford University Press.

Vaesen, Krist. 2008. A philosophical essay on artifacts and norms. Eindhoven University of Technology; PhD Thesis, Eindhoven, The Netherlands.
Von Wright, Georg Henrik. 1963. *The varieties of goodness*. London: Routledge & Kegan Paul.
Wetzel, Linda. 2006. Types and tokens. In *The Stanford Encyclopedia of Philosophy (Summer 2006 Edition)*, ed. Edward N. Zalta. URL=<http://plato.stanford.edu/archives/sum2006/entries/types-tokens/>.
Wright, Larry. 1973. Functions. *The Philosophical Review* 82 (2):139–168.

Chapter 5
Engineering design

5.1 Engineering design and making technical artefacts[1]

Modern engineers may be considered to be the most recent incarnations of the idea of *homo faber*, which stresses the tool-making aspect of human agency.[2] They have elevated the making of tools and of technical artefacts in general from craftsmanship to a science-based activity. Due to the increasing complexity of technical artefacts and the technical systems in which they are embedded, the making and (mass-)production of technical artefacts has evolved from experience-based crafts to professions where use is made of highly specialised scientific and technological knowledge. With this professionalization has come a division of labour that brings to the fore the different kinds of activities that often remain implicit and inextricably intertwined in the crafts. One of the main divisions of labour follows the dividing line between the mental and physical activities involved in making technical artefacts, between conceiving a technical artefact and actually making or producing one (Dym 1994, p. 15; Pahl and Beitz 1996).[3] The conceiving side is termed designing and it is done by specialised professionals, namely designers or designing engineers.

[1] This chapter is a modified version of Kroes (2009a).

[2] For an interesting discussion on the notion of *homo faber*, see Arendt (1958).

[3] For instance, Pahl and Beitz (1996, p. 1) remark that "The mental creation of a new product is the task of the design or development engineers, whereas its physical realization is the responsibility of manufacturing engineers."

P. Kroes, *Technical Artefacts: Creations of Mind and Matter: A Philosophy of Engineering Design*, Philosophy of Engineering and Technology 6,
DOI 10.1007/978-94-007-3940-6_5, © Springer Science+Business Media B.V. 2012

It is interesting to observe that this institutionalised division of labour between the mental and physical work involved in making technical artefacts runs parallel to the two kinds of activities that play a role in the way I have defined being an instance of a technical artefact kind. Let us recall that definition:

> Object x has the functional property of being a φ-er (or: is an instance of the technical artefact kind φ-er) iff x is the result of a largely successful execution of a largely correct design of a φ-er.

According to this definition the mental activity involved in making a technical artefact of a particular kind consists of coming up with a largely correct design. The physical activity is referred to as "a largely successful execution" of the design, that is, a largely successful material realization of the design. The definition suggests that the making of a technical artefact involves two steps, one involving mental and the other physical activity, and that these steps may be performed independently of each other and may therefore be separated in time. Each step has its own success criterion; the first mental step is successful if it produces a largely correct design and the following physical step is successful if it comes up with a physical object that is a largely correct execution of that design. The modern institutionalization of the division of labour between the mental and physical work involved in making technical artefacts is also based upon the assumption that these two activities can be pursued successfully more or less independently of each other. That assumption may be questioned; I will come back to this point shortly.

Because of the division of labour in making technical artefacts the outcome of a design process is not yet a technical artefact; what comes out of it is a design of a technical artefact. For the time being, a design may be taken to be, roughly, a description (which may include drawings) of a technical artefact. As such, a design is merely some kind of representation of a technical artefact (kind). A design may also include a plan or a description of how to make the artefact[4] in question and it may go on to function as a blueprint for its physical realization, that is, for the actual manipulation of matter so that it results in a particular kind of physical object. It is in this making phase that the production or manufacturing engineers and the production facilities come into play. Note that the designing aspect of making technical artefacts is only of particular interest in cases where making involves more than simply producing more instances or copies of an already existing kind of technical artefact. Copying the design of a technical artefact is not really designing.

Making designs or plans for new technical artefacts points to a feature of human agency that extends far beyond the domain of material production. We make all kinds of plans in the sense of considered series of actions, which may or may not involve the

[4] The notion of making a technical artefact is ambiguous; it can refer to the intentional creation (designing) of an artefact with a particular function or to its actual physical/causal production (for instance involving workers in a production facility who may not know what they are producing); for an interesting discussion on these two interpretations, see Thomasson (2007); here I concentrate on making technical artefacts in the first sense.

use or making of technical artefacts. The *homo faber* idea presupposes that human beings are *planning* agents, agents with the ability to form and execute plans (Bratman 1987). According to Bratman we are planning agents because we are, at least to some extent, rational agents in that we reflect on the outcomes of possible courses of actions, in other words, on the execution of possible plans. On top of that we need to plan our actions in order to coordinate our own actions in relation to the different goals we pursue simultaneously while coordinating our actions with those of others. The making of technical artefacts presupposes such planning capacity not only with regard to the production of an actual artefact on the basis of a plan or design but also with regard to the planning of how to use the technical artefact in question. In section 5.5 I will discuss in more detail a proposal to interpret engineering design as an activity of making 'use-plans' that may or may not involve the design of material means, that is, technical artefacts, necessary to realize the goals of these use-plans.

In order to avoid misunderstanding it is necessary to qualify the above-mentioned sharp distinction between the designing and the producing of technical artefacts in terms of mental and physical activities. Especially when it comes to mass produced products, the designing of a technical artefact may also involve actually making a prototype. The function of such a prototype is to test and evaluate the proposed design before it goes into mass production. In many engineering-design practices demonstrating that a proposed design 'works' by building a prototype is seen as an integral part of the actual designing phase, in particular of showing the correctness of a design. In such cases, the actual mass production of the technical artefact remains external to the design task but the designing part itself is no longer a purely mental activity since it involves building and running experimental tests on prototypes. Still, the outcome of the design phase, insofar as it is a plan for a technical artefact, is a mental product. In some design practices prototype creation may be virtually impossible (for instance, when designing a new harbour). Even then drawing a sharp distinction between design and production may be problematic because during implementation it may well be necessary to redesign part of the original design so that the design activity actually extends into the production phase. So, while conceptually we may distinguish between the mental and physical creation of technical artefacts it may in actual technological practice be difficult to separate them, just as this is the case with a craftsman making a technical artefact.

In this chapter I will restrict myself to the design aspect of creating or making technical artefacts. This is generally considered to be the most interesting aspect because it is assumed to require a certain degree of intelligence, inventiveness and creativity, whereas the production side is taken to involve only the actual material realization of a design whilst tapping the appropriate physical and organisational skills and making use of production facilities. This attitude towards the design and production aspects of technical artefacts may reflect the rather pervasive difference in the way science and technology (or more generally, mental and manual work) have been appraised as part of Western thinking since Greek antiquity. I will not question this attitude here but will merely reiterate my observation to the effect that in actually making technical artefacts both aspects may be intertwined in inextricable ways and recall Edison's famous saying to the effect that inventing technical artefacts involves one per cent inspiration (ideas) and ninety nine per cent transpiration.

Furthermore, I focus on the technological aspects of designing and not so much on its artistic aspects in which aesthetic features play a dominant role. Clearly, in some engineering design practices, such as in industrial and architectural design, aesthetic considerations may play a prominent part. There the notions of designing and design are often primarily associated with the aesthetic qualities of designed objects and aesthetic criteria may figure prominently in the criteria used to evaluate proposed designs. Such engineering design practices may come very close to art. A clear sign that we are dealing here with a different kind of design practice is the fact that the notion of correctness of a design becomes problematic and that it may become more difficult to reach consensus about when a design is (largely) correct. In many branches of engineering design, though, aesthetic considerations only play a minor role, if at all. There it is the design and development of technical artefacts that can fulfil practical functions that takes centre stage. In those branches, proposed design solutions are primarily evaluated on the basis of criteria such as effectiveness, efficiency, costs and durability rather than on aesthetic criteria. It is this kind of engineering design, in which the solution of technical problems has a major role to play, that I am primarily interested in here.

There is one preliminary issue that has to be addressed briefly before we can continue. Just a brief look at the range of engineering-design disciplines and at the divergence in the kinds of things designed by engineers is sufficient to raise the question whether it is indeed sensible to endeavour to generally characterise engineering design. There are dozens of different engineering disciplines (mechanical, electrical, civil, chemical, agricultural, bio(medical), material, mining, computer etc.) that design a myriad of products. Accordingly there is also great variety in engineering design practices, not only as far as the required competences and skills of design engineers goes but also in the composition of the design teams. Some design projects may be carried out by a single designer while others require large, multidisciplinary teams of design engineers. There is also much variety in the kinds of design problems that need to be solved. Vincenti (1990), for instance, distinguishes between normal and radical design problems and between design tasks that are high and low in the design hierarchy.

We have to ask ourselves whether there is any unity in such variety in design practices. Is it possible to generally characterise engineering design processes and to pinpoint domain-independent principles and procedures for engineering design? This has been a topic of controversy (see, for instance, Reymen (2001)). Naturally much depends on the level of abstraction chosen. It is easy to cite very general problem solving strategies (e.g. analyse, synthesise and evaluate) to characterise engineering design but in that way much if not all of what makes engineering design a specific kind of problem-solving is lost. Conversely, if we zoom in too much it becomes difficult to recognise the common elements in different design practices. One factor driving the search in engineering practice for systematic, domain-independent design principles is the ever-growing complexity of the objects of design. In recent decades this has led to the emergence of new fields of engineering such as systems engineering and design methodology which study the principles and procedures of engineering design in order to rationalise and improve design practice

(Cross 1994 (1989); Sage 1992; Pahl and Beitz 1996; Hubka and Eder 1996; Dym and Little 2000). Within these fields, the analyses of and proposals for methods of engineering design are often domain-independent. In the following section I will take as my starting point some general, domain-independent features of engineering design as proposed by engineers themselves.

5.2 Engineering design and science

Modern engineering design is a science-based activity but that does not make it a branch of applied science. Indeed, the solving of design problems is generally taken to be something different from the solving of scientific problems; a good designer is not *ipso facto* a good scientist or vice versa. Designing is even considered by some to be the salient feature of technology that distinguishes it from science (Mitcham 1994, p. 220). I will discuss two features of engineering design that make it an activity intrinsically different from scientific research. The first concerns the *decisional* nature of engineering design, the second the *wide variety of constraints* that have to be taken into account in designing.

As my starting point I take the following general characterisation of engineering design by the Accreditation Board for Engineering and Technology (ABET); it states that engineering design: [5]

> …is the process of devising a system, component, or process to meet desired needs. It is a decision-making process (often iterative), in which the basic science and mathematics and engineering sciences are applied to convert resources optimally to meet a stated objective. Among the fundamental elements of the design process are the establishment of objectives and criteria, synthesis, analysis, construction, testing and evaluation.

Although not explicitly stated, the systems, components and processes devised are assumed to be of a material nature; the design of (part of) an organisation or institution is not considered to be the domain of engineering design proper.[6] The stated objective of a design task is laid down in what is usually termed the list of specifications or the list of requirements. This list is derived from the (kind-proper) function that the object of design (which may be a system, component or process) is expected to satisfy.[7] That function, in turn, is related to certain human ends (or needs). If the designed artefact meets all the specifications, it is deemed suitable to

[5] http://www.me.unlv.edu/Undergraduate/coursenotes/meg497/ABETdefinition.htm; accessed November 14, 2006. Note the prevalence of the technology-is-applied-science idea in this conception of engineering design: the application of scientific knowledge to engineering design is explicitly mentioned. I take it that the 'desired needs' referred to in this quote stands for 'desires and needs'.

[6] According to Simon (1996 (1969), p. 111), however, the intellectual activity of designing material artefacts is not fundamentally different from the designing of organisational structures or procedures.

[7] In this chapter the term 'function' refers to 'kind-proper function'.

realize the desired function. Whether that indeed turns out to be the case depends very much on whether the list of specifications adequately captures the desired function. If it does and if the reasoning from end to function has also been performed adequately, then the use of the designed artefact may be expected to be a reliable means to help bring about the specified end.

A striking feature of the ABET definition is that it characterizes engineering design as a *decision*-making process and not simply, as is so commonly done, as a problem solving process. Characterising engineering design as mere problem-solving may indeed be misleading because of the dominant view that problem-solving involves finding or discovering the 'right' solution, the solution which, in principle, is uniquely determined by the problem structure. This discovery-picture has been traditionally associated with the kind of problem-solving that takes place in science or mathematics but it does not correspond to the kind of problem-solving peculiar to engineering design. Design problems are often ill-structured (Simon 1984) or wicked problems (Rittel and Webber 1984), which for instance means that there may be no definite formulation of the design problem itself, insufficient criteria to evaluate proposed solutions and no clear idea of the solution space. As indicated by the ABET definition, engineering design is partly about clearly establishing objectives and criteria by which to judge the proposed alternatives. Decisions thus have to be made that are to a large extent underdetermined by the problem formulation. Such decisions may have a significant effect on the aim and the outcome of the design project. But even if there is a clear and unambiguous formulation of the list of specifications, decisions have to be made concerning the promising options to work on. It may turn out that some of the specifications conflict, in which case decisions about trade-offs have to be made. Alternatively, given the state-of-the-art technology or the available resources it may not be possible to come up with a solution that satisfies all specifications. One then has to decide how the list is to be adjusted. When there is a set of alternative solutions satisfying all the specifications, there is no guarantee that one particular solution can be embraced as the best or optimal solution for rather fundamental reasons linked to multiple-criteria evaluations (see Franssen (2005); see also Kroes, Franssen et al. (2009)). In such cases the lack of rational procedures for determining the best option means that decisions again have to be taken about which option to choose.

These various kinds of decisions are all part and parcel of engineering design practice. The actual decisions taken may have far-reaching consequences for the outcome as they will shape the artefact that is being designed. This 'decisional' nature of engineering design reflects the idea that engineering design is much more a process of invention than a process of discovery. It is about the *creation* of new objects, not the discovery of what already exists. So the decisional nature of engineering design is not so much to be interpreted as the making of choices between pre-given, existing options but as the creating of various design options by making decisions about their properties.

Note that not all the problems that have to be solved in engineering design practice necessarily involve the kind of decisions discussed above; for instance, part of a design process may involve 'finding out' the maximum load a proposed construction can bear. Such problems are the domain of technological (engineering) research; they do not require the kind of 'creative' decisions that have to be made when solving design problems.[8]

The second important feature of engineering design to be discussed here is the variety in the kinds of constraints that design engineers have to deal with when designing. According to the ABET engineering students have to learn to solve engineering design problems under a (*ibidem*) "variety of realistic constraints, such as economic factors, safety, reliability, aesthetics, ethics and social impact." These various constraints are related to different values that have to be taken into account in designing technical artefacts. This variety of constraints is reflected in the list of specifications, which means that various factors (economic, safety, reliability, ethical etc.) co-define the design problem and therefore co-determine the ultimate shape of the object of design. Any proposed design will be assessed partly on the basis of how well it satisfies this variety of constraints. With scientific research problems most of the constraints mentioned are virtually absent. The outcomes of scientific research are assessed mainly on the basis of their cognitive values and because of this scientific problems are different from engineering-design problems.

Note that constraints on the object of design have to be carefully distinguished from constraints on design projects. Any design project has to be performed under constraints of time and resources. These project constraints are operative at the process level, but nevertheless they may also affect the outcome of a design project, but only indirectly, because they affect the means available for solving a design problem. The different role of constraints on the object of design and on the process of design in shaping the outcome of a design project becomes immediately clear from the fact that, once the constraints on the object of design and on the design process have been fixed, proposed design solutions are only assessed relatively to the constraints on the objects of design. From the point of view of process constraints there appears to be no real difference between engineering design and scientific research projects. The latter are also subject to all kinds of constraints, for instance, constraints deriving from limited resources, the risks associated with performing experiments, the possible social consequences etc. Again, these constraints may have far-reaching effects on the outcome of scientific research projects, because they have impact on the means for dealing with research questions. However, the outcome of scientific research questions is not assessed against these project constraints and therefore not shaped by these constraints.

Thus, engineering design and scientific research, as problem-solving activities, are governed by different kinds of values, norms and success criteria. According to

[8] This is not to say that science is not a creative enterprise. In science the creative aspect is traditionally considered to reside primarily in the activity of representing some pre-existing world, not in creating that world. This traditional view of science has come under attack from social constructivist quarters (see, for instance, Barnes et al. (1996)). Hacking (1983) has also challenged this view by claiming that in experiments physical phenomena are created. For a criticism of this view and an analysis of the differences between creating physical phenomena and creating technical artefacts, see Kroes (2003).

the 'ivory tower' model, science as a cognitive activity is ideally guided by epistemic values only, such as truth, empirical adequacy, simplicity and explanatory power. In practice this ideal may not be achieved but it nevertheless highlights an important feature of scientific research, namely that from a cognitive point of view the results ought to be basically independent of values prevailing in the wider social context. More or less the same may be said of engineering research or 'applied science': once the technologically interesting topics have been chosen, the research will proceed according to the same values as those abided by in science. Research, whether scientific or technological, is mainly guided by the values and norms of theoretical rationality that deals with issues of what to believe.

In contrast to science, the wider social context is of paramount importance to engineering design, since it is embedded within a broader framework of product creation processes. As part of these processes, problem solving in engineering design is subject to other kinds of values, norms and success criteria. Proposed solutions are evaluated in terms of pragmatic criteria such as effectiveness, efficiency, feasibility, costs and safety. Indeed, engineering design is guided by the demands of practical rationality which deals with issues relating to the course of action to take in order to achieve given ends. These kinds of actions and ends are always embedded in broader, value-laden social contexts that impose their own constraints on viable solutions. These constraints, moreover, are also subject to change in the course of time; for instance, in recent decades sustainability has emerged as a new important constraint on engineering design.

The following domain-independent definition of engineering design by Dym (1994, p. 17) suggests that assessing the quality of a proposed design involves two different kinds of criteria[9]:

> Engineering design is the systematic, intelligent generation and evaluation of specifications for artifacts whose form and function achieve stated objectives and satisfy specified constraints.

One set of criteria for evaluating proposed design solutions is the list of specifications (stated objectives). The other set consists of constraints that have to be 'satisfied'. Dym remarks that one may question whether a clear distinction between the set of constraints and the list of specifications can be made. For instance, the condition that a car engine has to satisfy a legal standard for pollution may be taken as an element of the list of specifications but it may also be seen as a constraint. Whether the distinction is meaningful or not, it is the variety in kinds of constraints or specifications imposed on design solutions that is important. Because of this variety conflicts between constraints/specifications may easily arise (for instance, between safety (more mass) and sustainability (less mass) requirements for cars). This is the

---

[9] Note that what Dym refers to as a specification for an artefact comes close to what I have been referring to as a design of an artefact; so his notion of specification is different from the notion of specification I have been using in expressions as 'the list of specifications."

reason why finding ‘clever’ trade-offs between conflicting specifications/constraints plays such a prominent role in engineering-design practice.

Clearly, modern technology is strongly science-based not only in the sense that it makes use of scientific knowledge but also in the sense that technological research is guided by scientific methods. This is also true for engineering-design practice. Nevertheless it would be a great mistake to consider problem-solving in engineering design to be a kind of applied science. Admittedly, a lot of problem-solving that goes on in engineering-design practice may aptly be characterized as ‘applied science’ or ‘engineering science’. This is the problem-solving that is research-oriented, and as I remarked already, it is not very different from the kind of problem-solving found in science. But that is not the kind of problem-solving that I am interested in here since it is not characteristic for solving *design* problems. That kind of problem-solving is very much distinct from the kind of problem-solving in science because of the two features mentioned above, its decisional nature and the wide variety of constraints. Below, in section 5.4, I will discuss yet another feature that distinguishes engineering design from scientific research. It concerns the kind of reasoning employed by scientists and design engineers in their problem-solving. Virtually all analyses of scientific reasoning construe it as variations of inductive, (hypothetico-) deductive or abductive reasoning. In designing, a different kind of reasoning takes centre stage, namely means-end reasoning. First, however, I will have a closer look at what kind of activity engineering design is. For that I turn, for a second time, to Simon’s *The sciences of the artificial* (1996 (1969)) in which he presented not only an interesting analysis of technical artefacts but also of designing.

5.3 Engineering design: from function to structure

According to Simon engineering design deals with the synthesis of artificial things and engineers, in particular designing engineers, are (1996 (1969), p. 4–5) “concerned with how things *ought* to be – how they ought to be in order to *attain goals*, and to *function*.” Instead of taking the world for what it is (as in science) engineering design seeks to change the world to meet given needs, desires or goals. Whereas in science our ideas and beliefs are adjusted to how things are in the world, the engineering attitude is precisely the opposite, namely to adapt the world to our ideas, desires and needs. This difference in attitude between science and engineering may be expressed by the difference between a “mind-to-world fit” and a “world-to-mind fit.” More specifically, engineering design contributes to the development of the material means that people may use to achieve their goals that involve changing their physical environment. These material means, technical artefacts, have a function; when functioning and when used properly they are supposed to bring about effects that are conducive to achieving the ends associated with their function. The normative character of what engineers have to deal with is reflected in normative statements about technical artefacts. In the previous chapter the normativity of technical artefacts has been analysed in detail (section 4.1), which led us to the conclusion that this normativity is not intrinsic to technical artefacts but derives from the normativity of human action.

As we have seen, Simon distinguishes between the inner and outer environment of technical artefacts. The outer environment is very important because it is what moulds the artefact. He considers a technical artefact to be a kind of (1996 (1969), p. 6) 'interface' between the inner and outer environment, the surroundings in which it operates." The inner environment of the artefact, its physical make-up, is shaped in such a way that it realises the goals set in the outer environment. Therefore engineering design, more generally the sciences of the artificial, must focus, according to Simon, on this interface, since the (1996 (1969), p. 113) "artificial world is centred precisely on this interface between the inner and outer environments; it is concerned with attaining goals by adapting the former to the latter."[10] The task of engineering design is to come up with descriptions of technical artefacts for which the inner environment is appropriate or adapted to the outer environment.[11] This interface character of technical artefacts explains the difficulties engineers have when disambiguating and fixing the meaning of the notion of function, especially in relation to notions of physical behaviour (capacities) and purpose (see 2.5.1).

One of the reasons why Simon's analysis of technical artefacts and engineering design is so interesting is that it draws attention to the tensions between the inner and outer environments of technical artefacts, the tension between what artefacts do or are capable of doing, and what they are expected to do within some context of human action (the 'rich' outer environment that imposes so many constraints). It is this tension which, according to Staudenmaier (1985, p. 103), is the defining nature of technology. Indeed, engineering design is about filling in the "substance and organisation" of the inner environment so that it meets all the requirements or constraints imposed from the outer environment. In so doing, engineers have to take into account what is physically and technologically possible. It is the tension between the set of physical and technological constraints that apply to the contents of the inner environment and the set of constraints that derive from the outer environment (contextual constraints: functional specifications and other requirements) that defines the core of engineering design (see Fig. 5.1) (for more details, see Kroes (1996)). This tension is one of the main driving forces behind the development of technical artefacts (see, for instance, Petroski's (1992) principle of 'form follows failure'). Obviously apart from these 'market pull' factors, advances in technology may also drive the development of technical artefacts (the 'technology push').

If we take engineering design to be a process of devising a technical artefact that is adapted to some specific environment, then in Simon's terminology it starts from the outer and proceeds to the inner environment. But as we have seen in chapter 2,

---

[10] This remark suggests that there is just a one-way influence from the outer to the inner environment. The design of technical artefacts, however, may also be a matter of adapting the outer to the inner environment (for instance, by adapting the behaviour of prospective users through training).

[11] It is the distinction between inner and outer environment that also lies at the basis of Hubka and Eder's (1996, p. 108-114) theory of the properties of technical systems (technical artefacts).

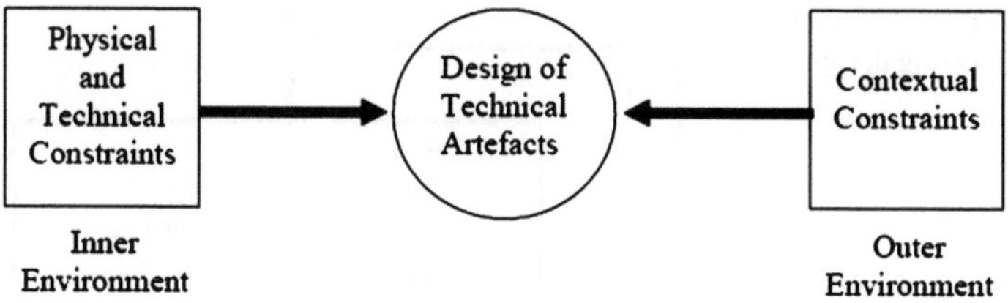

Fig. 5.1 Design and the tension between inner and outer environment

different kinds of descriptions of a technical artefact are associated with the outer and inner environments, respectively a functional and structural description. From the point of view of the object of design, an engineering-design process may therefore be taken to start with (a description of) its function, its intended behaviour, and end with (the description of) a physical structure that realises that intended behaviour.[12] In other words, it is a process that runs from one of the constitutive elements involved in the dual nature of technical artefacts, from (a description of) human intentions, to the other one involved, to (a description of) a physical structure. One might say that technical artefacts inherit their dual nature from the way they come into existence.

Since the functional description with which a design process starts is a black-box description of the object of design (see section 2.4), designing may be described as opening up this black box and filling it with a physical structure that is able to perform the required function.[13] That physical structure has to be put together, synthesized, from parts (components) some of which may already exist, others of which may have to be tailor-made for the specific case at hand. It is for that reason that designing is often characterized as a *synthetic* activity, or as an activity that uses *synthetic methods*, as opposed to scientific research that is commonly characterised as an *analytic* activity, or as an activity that uses *analytic methods*. However appealing this characterization of design and research may sound, it is not so easy to explicate the precise difference involved. Scientific research also involves synthetic activities; researchers also have to be skilful designers, not of technical artefacts or services but of theories, experiments and the equipment needed to perform such experiments. From a philosophical point of view little is known about what are the distinctive synthetic features of engineering design, if any, that make it different from synthetic activities in other fields, particularly in science. Though the intuitive characterisation of designing as a synthetic activity, as opposed to research as an analytical activity, may be very attractive, this specific conception of the synthetic-analytic distinction is in need of further clarification. It may well be that just as the distinction between analytic and synthetic *statements* has come under attack (Quine 1951), so the distinction between analytic and synthetic *methods* will have to be reconsidered (for a discussion on this distinction see, for instance, (Beaney 2007)).

[12] See also Dym and Little (2000, p. 113).

[13] In actual design practice most design problems concern variations on existing technical artefacts; then, at the beginning of the design process most of the content of the black box is already known.

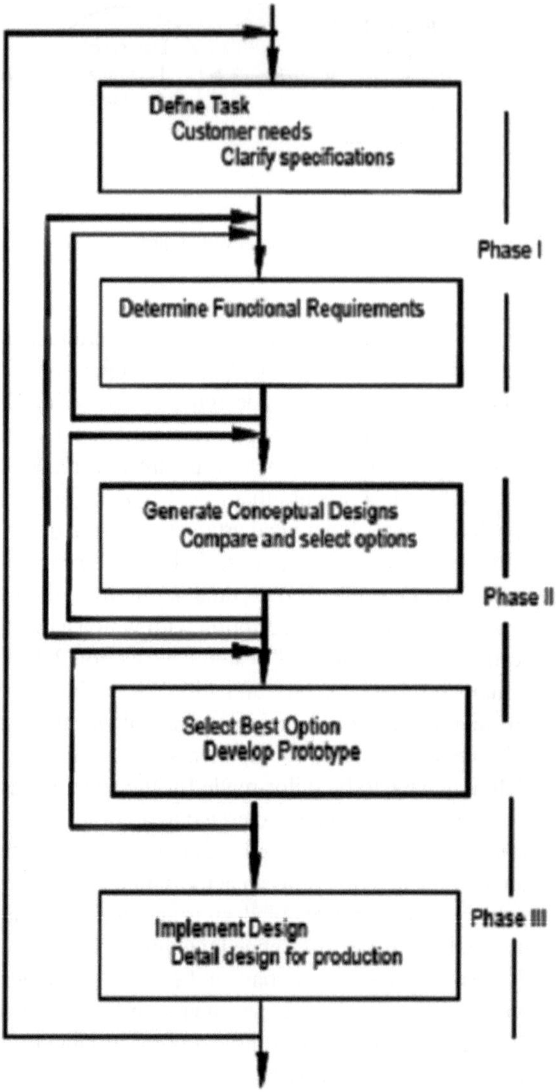

Fig. 5.2 A phase diagram of the engineering design process

So a design process may be characterized, for short, as a process in which a transition is made from a function to a structure. How is such a transition to be accomplished? That has been and still is a matter of great concern to design methodologists. They have come up with various models of design processes in terms of rationally prescribed steps or phases and have developed many design tools intended to support designers in their work. These phase models and design tools are supposed to contribute to the improvement of actual design practices. As the example given in Fig. 5.2 (Kroes et al. 2009, p. 587) illustrates, most of the models are more or less detailed variations on the basic analysis-synthesis-evaluation cycle.[14] As long as designing remains an activity performed by one single individual, these phases will be mainly relevant from a conceptual point of view. As soon as designing becomes a matter of teamwork, however, which tends to be the situation in modern industry where complex and large systems are dealt with, the phasing of the design process becomes an important management tool for the division of labour and for organising, controlling and steering the process of product development.

One matter that hampers discussions on the usefulness of implementing such phase diagrams in engineering practice are the criteria for evaluating and measuring the success of the outcome of an engineering design process. From a strictly engineering point of view, the simplest success criterion is to meet the list of specifications while satisfying the given constraints. This assumes that the list of specifications is immutably fixed at the beginning of the design process, which is not often the case. Because of problems encountered on the way, they may have to be adjusted during the design process. Moreover, as was remarked before, decisions about which performance criteria to use and the development of methods for measuring such performance criteria are often an integral part of the design process. On top of this, various participants in the design process may evaluate the outcome in different ways. In spite of these difficulties, design methodologists claim that the implementation of systematic approaches to design improves the design process (see, for instance, Pahl and Beitz (1996, p. 499–501)).

There is one design tool, namely the method of functional modelling or functional decomposition, which is worth mentioning here.[15] It is a standard tool of design engineers that supports them in making the transition from function to structure. It is a kind of 'divide and conquer' strategy in the sense that it assists engineers in splitting up the overall function of the object to be designed into sub-functions, each of which in turn may be further subdivided into sub-sub-functions, and so on (for an example, see Fig. 5.3). Usually, this process stops the moment a standard solution for one of the sub-functions is available. For that sub-function it is then easy to make the transition from structure to function: a physical structure that realizes the sub-function involved is already known. It may arrive, however, that this is not the case for a particular sub-function and that it is also not clear how to further subdivide it. How is, in that situation, the transition from function to structure to be made? One way to deal with this question is to make an appeal to the 'creativity' or the 'inventiveness' of the designer and leave it to that. However, such an answer is not really satisfying. It blocks any way of explaining how a designer came to a certain design; 'creativity' and 'inventiveness' become magical assets of designing engineers. Such an answer, moreover, appears to be not in line with engineering practice in the sense that designing engineers in fact often reason from function to structure and the other way around. Once a designer has come up

---

[14] See, for instance, Cross (1994 (1989), Ch. 2).

[15] For a discussion of different forms of functional decomposition in engineering design, see Van Eck (2011).

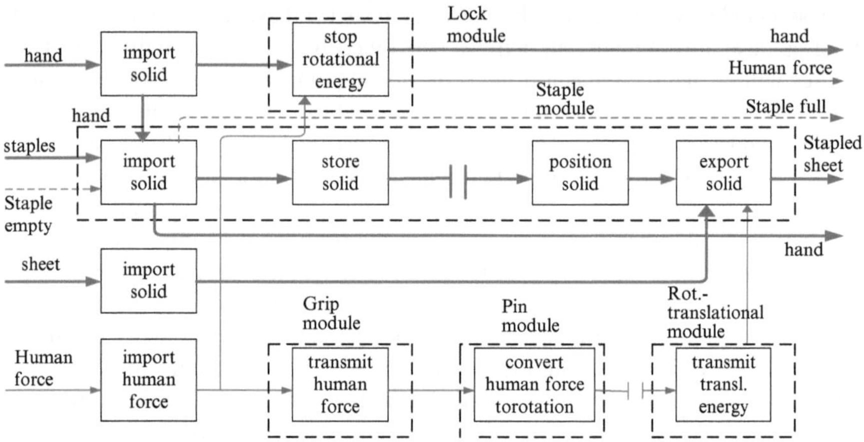

Fig. 5.3 Example of a functional model of a stapler (taken from Van Eck (2011, p. 144))

with a (brilliant) idea for a design, the problem remains of justifying that design solution, showing that it will work (which may involve explaining why that particular structure realizes the function). In making the transition from function to structure more appears to be involved in engineering design practice than 'brute' forms of creativity and inventiveness.[16]

Apart from pragmatic issues about how the transition from function to structure may be supported in design practice, this transition raises also philosophical issues about what kind of reasoning may connect functions to structure and what kind of knowledge this involves.[17] There appears to be, as I have observed already before (see section 2.4), a logical gap between the functional and structural descriptions of technical artefacts. Nevertheless, the function and structure of technical artefacts are taken to be intimately related, not only in the sense that the physical structure realizes (or is supposed to realize) the function, but they also constrain each other (not any object can perform any function). Moreover, designers are apparently able to reason successfully from functional to structural descriptions or vice versa, (for instance when they justify a proposed design by explaining why it realizes the required function).[18] What kind of reasoning might be involved in translating a function into a structure in engineering design?

[16] For a philosophical analysis of functional decomposition and its relation to the problem of making the transition from function to structure in design, see De Ridder (2007, chapter 5).

[17] I am not so much interested in how from a context of discovery perspective design engineers succeed in finding solutions to design problems (what kind of heuristics they use, how they deal with ill-defined problems, with uncertainty et cetera) but how from a context of justification perspective design engineers justify, when challenged, their design choices and the design solutions they come up with. What kind of reasoning do they employ then? So, my question about how they reason from function to structure is to be understood, not in a historical, but in a rational reconstruction sense (Kroes et al. 2009, p. 568-569).

[18] For a discussion on the 'coherence' of structure and function of technical artefacts, see Kroes (2006).

5.4 Means-end reasoning

In practice designers make use of methods like constructing morphological charts or function-means trees to go from function to structure; these charts or trees give a graphical representation of functions (and sub-functions) and the various known ways of realising them (Dym and Little 2000, p. 116, 146 ff; Cross 1994 (1989), p. 106 ff). They present in a condensed way the available alternatives for filling in the black boxes corresponding to the (sub)functions. As mnemonic devices they do not give us any clues about what kind of reasoning may lead from a function to a structure. It may be expected that 'means-end' reasoning plays an important role in translating a function into structure, since the design process is all about finding or constructing the appropriate means for achieving certain ends. In spite of its apparent importance to engineering practice and daily life in general, the formal (logical) analysis of means-end relations and reasoning has received relatively little attention (Von Wright 1963, 1972; Segerberg 1992). Recently, research into artificial intelligence has triggered more interest in this kind of reasoning (Pollock 2002). Within the practice of engineering design there is also a great interest in the formal analysis of functional reasoning, a phenomenon which appears to be closely related to means-end reasoning, because of attempts to develop formal tools to represent the objects of design and supporting functional reasoning about these objects (Chittaro and Kumar 1998; Dym 1994).[19]

Means-end reasoning may be seen as a form of practical inference about what needs to be done to achieve an end. In that respect actions are taken to be means to certain ends (states of affairs in the world) (Hughes et al. 2007). From a technological point of view, however, objects may also be viewed as means to achieve ends (for instance, a knife is a means for cutting bread, a pencil a means for writing). A formal analysis of means-end relations and reasoning in which objects and not actions are means, has still to be developed.

In his seminal paper on practical inferences Von Wright (1963, p. 161) analyses the following kind of reasoning pattern:

One wants to attain *x*.
Unless *y* is done, *x* will not be attained.
Therefore *y* must be done.

Von Wright calls *x* the end and *y*, which is an action, a means to that end. This type of argument concerns necessary means to ends and the conclusion, which states an action, expresses a practical necessity. This practical necessity is derived

[19] For a general discussion of means-ends reasoning, see Hughes (2009).

from the statement of an end and from a conditional statement based on the causal structure of the world. The question one has to ask is whether such arguments are logically conclusive. *Prima facie* this seems not to be the case since the premises consist of descriptive statements and the conclusion is prescriptive. But for Von Wright this is not a convincing argument against logical conclusiveness.

Two features of this type of arguments are of particular interest from the point of view of engineering design. The first has to do with the fact that in practical inferences a transition is made from descriptive to prescriptive statements, from what is the case to what must be done. This may be taken as a strong sign that this kind of reasoning is highly relevant for engineering design. However, the relation of practical inferences to reasoning from function to structure is an open matter. In particular it is not clear how practical inferences relate to the fact that reasoning from function to structure and vice versa is problematic because of the alleged logical gap between functional and structural descriptions of technical artefacts, a logical gap stemming from the is-ought dichotomy. The kind of practical inference studied by Von Wright concerns practical necessity in relation to human actions and its conclusion is phrased in terms of what "must" be done. But function statements about technical artefacts are connected to statement about what objects, when considered as means, should do or ought to do and not about what human agents must do. Furthermore, engineering design is primarily about reasoning from function to structure, that is, reasoning from the normative to the descriptive, whereas in practical inferences of the kind considered above the reasoning proceeds in more or less the opposite direction, from the descriptive to the prescriptive. Finally, given the multiple realizability of technical functions, practical necessity seems too much to ask of reasoning patterns from function to structure in engineering design.

The second feature is that the second premise, which is founded on a conditional relation between the means and the end, is based on a causal relationship. Not surprisingly this closely links means-end reasoning to the causal structure of the world. If we know that event A causes event B,[20] then we may realise the occurrence of event B by bringing about event A, if this is technologically possible and if there are no interfering circumstances. So the action of bringing about event A may be considered to be a *means* for the occurrence of event B, considered as the *end*. The causal relationship in itself does not imply practical necessity, that is to say, bringing about event A is not a necessary means for bringing about event B. For that to happen a much stronger conditional statement is required, namely that the bringing about of the occurrence of A is the only practically feasible course of action for bringing about event B.

The intimate relation that exists between means-end reasoning and causal relations explains why scientific knowledge plays such a dominant role in modern design practice. This leads to the question of the kind of knowledge used to solve design problems. As has been argued, it would be misleading to assert that

[20] More precisely, tokens of event type A cause tokens of event type B etc.

engineering design is simply the application of scientific knowledge (or knowledge produced by the engineering sciences). A knowledge of natural phenomena is certainly not all that is needed to solve design problems. Hubka and Eder (1996) have attempted to develop a design science, which they take to be a system of logically related knowledge about designing and for designing. In their enumeration of the various kinds of knowledge needed for engineering design, knowledge from the engineering sciences is just one item in a long list (1996, p. 72). In a similar vein, Dym and Little (2000, p. 22–23) remark that the majority of the many questions that have to be posed when designing a relatively simple object such as a safe ladder cannot be answered by applying the mathematical models of physics. According to Vincenti (1990, Ch. 7) the anatomy of engineering-design knowledge includes at least six different categories of knowledge, some of which do not derive from scientific knowledge at all, such as the 'know how' acquired on the shop-floor. All these various kinds of knowledge are important for turning a functional description of the object to be designed into a structural description. Ryle's (1984) distinction between 'knowing that' and 'knowing how' may be of particular relevance when analysing the kinds of knowledge used to solve design problems, because of the intimate relationship between designing and knowing how to make or do things.

To conclude, when compared to science the kind of problem-solving prevalent in engineering design not only appears to employ distinctive forms of reasoning but also distinctive forms of knowledge. Up until now, the nature of design knowledge, or more generally technological knowledge, has not received much attention in epistemology.[21] This is even more true of the formal analysis of means-end reasoning in logic.

5.5 Designing plans

So far I have analysed the nature of engineering design mainly from the point of view of the object of design, of the technical artefact to be. My overall perspective has been the making of a (new) technical artefact and I have particularly analysed how a technical artefact, as an object of design, is described at the beginning and at the end of the design process. This object-oriented view on engineering design is rather dominant among engineers. It is true that the usual characterisation of the outcome of a design process stresses that it is a production plan and not a description of a real material object, but this is simply a consequence of the prevailing division of labour. The design phase is followed by the production phase which results in the real, material technical artefact to which the designing was geared. If one changes perspective, though, from the design and manufacture of a technical artefact to its use one sees that this cannot be the whole story behind engineering

[21] For a discussion on the nature of technological knowledge, see Houkes (2009).

design. From the perspective of the user it is not the making of a technical artefact that matters but rather how to use it in order to realise the user's goals. To that end it is not a fabrication plan that the user needs but an instruction manual, which describes how the artefact is to be used. An instruction manual or use-plan is needed to make the function 'accessible' to the user. An unfamiliar technical artefact without a manual or a use-plan may be of no practical use.[22] Thus, from the point of view of the user it is not the production plan for the material technical artefact that matters, but rather its manual or use-plan.

It may well be the case that when designers characterise the outcome of design processes as production plans for technical artefacts, they implicitly take the manual to be part of the technical artefact. It is, however, important to make its role explicit because it enables attention to shift from material objects to actions and plans in which objects have a role. From an action-oriented view, engineering design may be interpreted as making use-plans that describe how goals may be realised with the help of technical artefacts. Technical artefacts may be said to be embedded within such use-plans. Following this line of reasoning, Houkes, Vermaas and others (2002; 2010) have developed an action-theoretical account of the designing and using of technical artefacts. In it they reconstruct the design and use of technical artefacts in terms of plans, intentions and practical reasoning. They take plans to be goal-directed series of considered actions and they see a use-plan for an object as a series of actions involving the manipulation of the object in order to achieve a given goal. They divide the design process into two different activities, namely use-plan design and artefact design. Each of these design activities is reconstructed in terms of plans and the plan for artefact design is embedded in the plan for use-plan design (see Fig. 5.4). In their account, the interaction between designers and users does not simply involve the transfer of a technical artefact but also, and primarily, the communication of a use-plan (Vermaas and Houkes 2006, p. 7).

An attractive feature of this action-theoretical interpretation of engineering design is the central role it attributes to practical rationality or practical reasoning. If plans are the outcome of engineering design then these plans, irrespective of whether they involve the manipulation of objects, have to satisfy the demands of practical rationality. This applies to the use-plan itself but also to the plan for artefact design which is embedded in the use-plan design. Using the work of Bratman (1987, p. 31), Houkes and Vermaas (2010, p. 37–41) discuss four demands to be placed on plans. The core demand concerns the *effectiveness* of plans; the correct execution of a plan should lead to the realization of its goal. Assessing whether that is indeed the case may be a rather context-sensitive matter, since it may depend on what the executer of the plan regards as a satisfactory result. Furthermore, plans should be *goal consistent*; they should not include incompatible goals. They should also be *belief consistent*; plans should be consistent with the justified or reasonable beliefs

[22] The most common form of a use-plan is a written manual; simple technical artefacts often come without a manual as the-use-plan is presumed to be known to the user and so remains implicit.

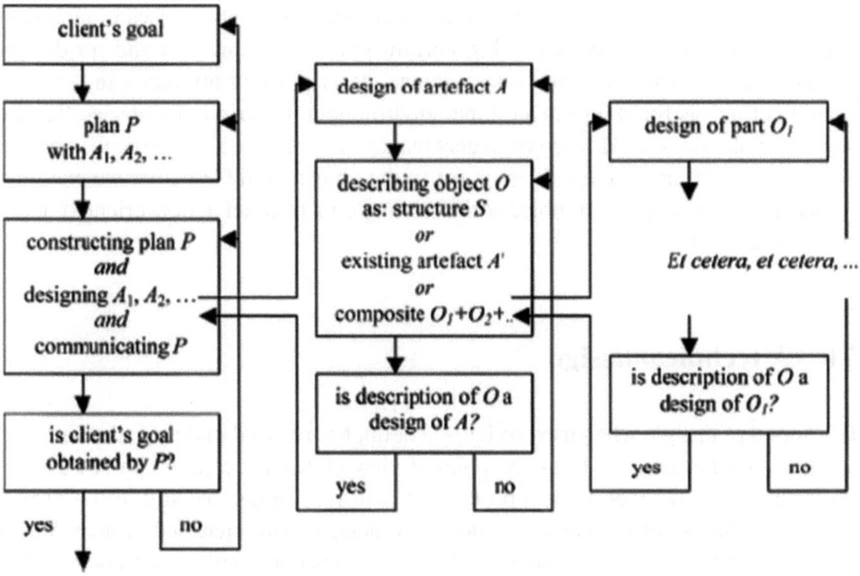

Fig. 5.4 Plan for artefact design embedded in plan for use-plan design; reprinted from Design Studies, 23(3), Wybo Houkes *et al.* 'Design and use as plans: an action-theoretical account', 303–320, 2002, with permission from Elsevier

of the agent who executes them. Finally, plans should be *means-end consistent* which requires that they can be broken down into preliminary steps, sub plans and means so that in the eyes of the agent they may be successfully executed. According to Houkes and Vermaas (2002, p. 320; 2010, pp. 41–44) these demands of practical rationality may lead to norms for good and bad design and use. They question, however, whether such demands on plans exhaust the norms operative in engineering design and artefact use. They note that their approach to engineering design and artefact use has an intellectual bias: in line with what was posited in section 5.1, the actual execution of plans is usually not considered to be an interesting topic in its own right. This leads to an interpretation of the demands placed on *practical* rationality that relates primarily to rational *deliberation*, a situation which also appears to be the case regarding the demands that Bratman imposes on plans. Actually making things or executing plans may impose additional demands. For instance, it is not clear whether or to what extent the above consistency constraints can account for the important role of the norm of *efficiency* in engineering design.

This action-theoretical approach to engineering design analyses the nature of designing and its output primarily from the point of view of what Simon calls the outer environment. It takes as its starting point practices of intentional human action in which technical artefacts are used to realise ends. Without recourse to this context of human action it is impossible to adequately characterise engineering design and technical artefacts. This approach therefore brings with it the same problem of how to translate a function into a structure as that encountered in the object-oriented approach to engineering design. From an action-theoretical point of view, technical artefacts provide ways of achieving certain goals; but how can and do design engineers move from an 'outer environment' description of artefact x in terms of what it is for (x is for φ-ing) to an 'inner environment' description that specifies the physical make-up of x? How do engineers manage to jump back, so to speak, over the 'for-operator' from what an object is for to the object itself? Whether we examine engineering design from an object-oriented angle or from an action-oriented angle this problem remains.

5.6 A technical design

The notion of design, when used so far as a noun, has referred mainly to the outcome of a given design process. From the point of view of product creation processes, this outcome is usually taken to be a production plan for objects that still lead a virtual existence. This is not the noun-type notion of design I am interested in here. When referring to a car design, for instance, what is meant is not usually its production plan but something that has more to do with the properties of the car itself, irrespective of whether that car actually exists or of how (if it indeed exists) it was actually produced. It is not easy to grasp what this 'something' is. Whatever it is, the design of the car is an important facet since it more or less determines the accompanying structural and functional properties. According to our conception of a technical artefact kind, it even is a defining feature of the car in the sense that its design makes the technical artefact an instance of a particular technical artefact kind, namely the kind 'car'.

In order to try to come to terms with what, in this sense, a design is, consider the design of the Newcomen steam engine, which is represented graphically in Fig. 5.5. The main function of this kind of steam engine was to power pumps to drain mines, and this was achieved by producing a reciprocating motion in the great beam, which was activated by the motion of the piston, and so on. The drawing provides some information about structural features of the design of a Newcomen engine: about some of the parts it is made of, their form and layout. Even if we were to add all the relevant structural information to this drawing we would still not end up with a full representation of the design of the engine. As a designed object, the Newcomen engine has a function, which is intimately related to human purposes, but that function is not contained in the structural representation of the design. For a representation of this aspect of the design of the Newcomen engine it is necessary to add information about its overall function, the functions of its parts, means-end relations and how the machine operates. In order to highlight the intimate relation of a design to human purposes, a representation of a design has to include information about its structural and functional features.

Note that, by contrast, a design as a production plan for a still virtual technical artefact does not necessarily include information about the functional properties of that artefact and all its parts. According to Dym and Little (2000, p. 10) a production plan has to be such that "the fabrication specifications must, on their own, make it

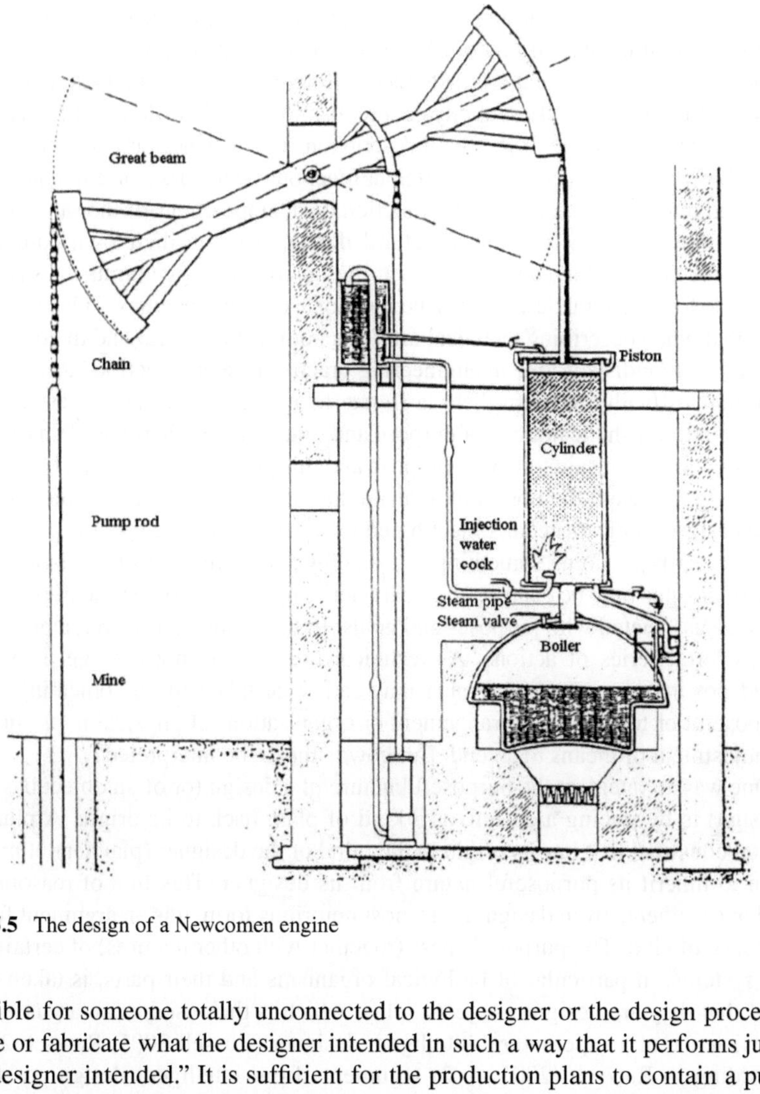

Fig. 5.5 The design of a Newcomen engine

possible for someone totally unconnected to the designer or the design process to make or fabricate what the designer intended in such a way that it performs just as the designer intended." It is sufficient for the production plans to contain a purely structural description of the technical artefact. In principle it is not necessary to include a functional description of the artefact since a functional description does not specify in a "clear, unambiguous, complete, and transparent" (*ibidem*) way the physical properties of the object to be produced. So the notion of a design seen as a production plan is clearly different from the notion of a design that is central to determining the kind to which a technical artefact belongs.

Anything that is called a design of a technical artefact may vary greatly in engineering practice in terms of level of detail and can be anything from a rough sketch, as displayed in Leonardo Da Vinci's drawings of machines, to a complete description of every minute detail of a prospective or existing artefact. I will assume that a design as a representation of a technical artefact has to be a combined description of all of its relevant physical and functional properties (relevant in the light of the performance of its overall function). A functional description represents only half of the design of a technical artefact since different physical structures may realise the same function and different physical realisations imply different designs of the artefact. The same is true of structural descriptions of a technical artefact: one and the same physical object may perform different functions on the basis of different designs (reflected in different structural and functional decompositions of the same object). Neither the functional nor the structural descriptions on their own completely capture the design of a technical artefact; the functional design omits the structural side while the structural design lacks the functional design properties. This just goes to show that when describing technical artefacts both the structural and the functional properties are indispensable in engineering practice (see also section 2.4).

A main difficulty when further clarifying the notion of a technical design lies in its association with the notions of purpose and function. Artefacts based on a technical design are often said to have a purpose and this purpose is conferred on them by their design. Indeed, the notion of a design has strong teleological connotations in that a designed object (i.e. an object based on a design) has the specific property of '*for*-ness': it is *for* doing something or *for* being something. This teleological character of designs may be captured by characterising them as some type of plan since plans are associated with purposes and goals. In this context, however, a plan is not a considered series of actions. As technical artefacts do not execute plans, that would not make any sense. A plan may rather be taken to be something like a 'purposeful or teleological arrangement or organisation' of physical parts showing the adjustment of means to an end. But how is this to be interpreted?

One way to interpret the purposeful nature of a design (or of an object based on a design) is by tracing a design, as a kind of plan, back to its origin. A plan is a mental construct that has its origin in the mind of the designer (planner). It may be taken to inherit its purposeful nature from its designer. This line of reasoning is used in arguments from design. In its most notorious form, it is an argument for the existence of God. The purposefulness (together with other features) of certain natural systems, in particular of biological organisms and their parts, is taken to be proof that they are designed objects, a fact which is then used as an argument for the existence of a supernatural intelligent designer (Russell 2005; Ratzsch 2005). According to Ratzsch (2005, p. 2) arguments from intelligent design are rather unproblematic in the case of technical artefacts or, more generally, in the case of things that "nature *could* not or *would* not produce." He claims for instance that for a DVD-player the conclusion that it was designed by human beings is "nearly inescapable".[23] A similar claim was made more than two hundred years ago by Paley with regard to a watch; he stated that when we examine a watch, what we see are (Paley 2006 (1802), p. 14):

> contrivance, design; an end, a purpose; means for the end, adaptation to the purpose. And the question, which irresistibly presses upon our thoughts, is, whence this contrivance and

[23] Fehér (1993) presents an interesting thought experiment that puts this claim to the test.

> design. The thing required is the intending mind, the adapting hand, the intelligence by which that hand was directed.

In this way, the purposefulness of a technical design (and of a technical artefact based on that same design) may be directly related to, and considered to be derived from, the intentionality of a human designer.

Still, this does not lead to a clearer picture of what a design as a defining feature of a prospective or real artefact is. Things become even more complicated when a technical artefact, as a designed object, is taken to be the 'embodiment' or 'material realisation' of a design. What does it mean for a physical object to embody a design, a mental plan, and to what extent does it inherit the purposefulness of a design? One way to interpret such characterizations of technical artefacts is along the lines of the dual-nature thesis I have presented in the preceding chapters. From that perspective a design is somehow intimately related to the structural and functional features of a technical artefact. A design has to refer to the structural features of technical artefacts in order to account for their causal efficacy and to their functional features to account for their purposefulness. Insofar technical artefacts, as physical objects based on a design, are products of the mind they inherit the teleological nature of the intentional action of the designer.

To conclude, from a conceptual point of view a clear analysis of the idea that technical artefacts embody a design is still lacking. In engineering practice these conceptual problems do not appear to be very important. In fact, in contrast to the vast amount of literature on the notion of design as an activity, one can search almost in vain in engineering handbooks for an elaborate analysis of what a design, in the sense intended here, incorporates.[24] From a pragmatic point of view, what is much more important is how designs of technical artefacts may be unambiguously represented. The growing complexity of modern technical artefacts and the use of computers in supporting solutions to engineering design problems have increased the need for more rigorous, formal representations of designs. Such representations are vital to the development of engineering data management systems and for computer aided design (CAD). It is especially the formal representation of functions that proves to be problematic (Dym 1994). Much work is currently being done on developing taxonomies of functional primitives (a field sometimes referred to as 'functional modelling'), on functional representation and functional reasoning with the aim to support engineers in their solving of design problems and in the accurate representation of designs (see for instance Wood (2009)). The main obstacle in these fields turns out to be the formal modelling of functional features of designs and technical artefacts.

[24] When used as a noun, the notion of design usually refers to a fabrication plan. Hubka and Eder (1996, p. ix) mention the interpretation of a design as the outward appearance and pattern of artefacts; this interpretation is not particularly of relevance to the present discussion. For a proposal of a general notion of design that covers not only technical but also biological designs and that is based on the notion of type fixation, see Krohs (2009).

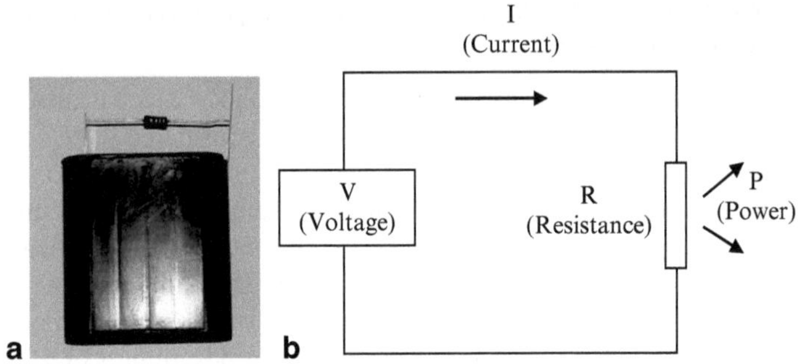

Fig. 5.6 A picture of a physical system (5.6a) and a graphical model of this system (5.6b)

5.7 Formal modelling of functional design features

The formal modelling of technical functions is of direct importance for engineering (design) practice; when successful it may support engineers in archiving and retrieval of functional knowledge about technical artefacts, in unambiguous communication about functional features and, since these are closely related to designer's intentions, about designer's intentions, and by making possible computer-assisted functional reasoning.[25] So far, however, attempts to develop formal representations of functions have not been very successful. In my opinion this is due to the fact that the functional features of technical artefacts are mind-dependent; they are always assigned by humans. As a consequence, the formalization of functional features of technical artefacts requires a formalization of the notion of function assignment. To illustrate this in more detail, I will compare the formal representation of a physical object with the formal representation of a technical artefact.

The formal representation of physical systems and their behaviour has met with ever greater success since the mathematization of physics in the sixteenth and seventeenth centuries. Nowadays it is possible, and even common practice, to represent (complex) physical systems and to simulate their behaviour on computers. As an illustration, consider the formal representation of the very simple physical system of Fig. 5.6a, an electric system consisting of an electric power source (with voltage V) and a resistance (R). Fig. 5.6b contains a graphical model of this physical system. This model is not yet a formal model of the system because from a formal point of view the language used to represent the system is still ill-defined.

A step further down the road toward a formal model consists of describing the electrodynamic behaviour of this system in a mathematical way. The behaviour of this system is governed by Ohm's law and if we take also into account that the

[25] For a more extensive treatment of the issue of formally modelling functions, see Kroes (2010).

power P dissipated by the resistance is the product of the voltage and electric current, then its behaviour is described by the following set of equations:

$V = I.R$

$P = V.I$

The combination of these equations is often taken to be by itself a formal model of this physical system, since it makes use of the formal language of mathematics. For our purposes, however, it will be more convenient to take a formal model (M) of a physical system to be in general an ordered n+m-tuple of the form:

M (physical system) = <V1,..Vn, R1,…Rm>

with V1,..Vn representing the physical variables and constants that characterize the state of the system (at a certain time) and with R1,…Rm representing relations between these physical variables and constants. The relations R1,…Rm may include laws of physics but also relations that find their origin in the specific make-up of the physical system involved.[26] When we fill in these relations for the electric system we get the following formal representation:

M(electric syst.) = <V, I, R, P, V = I.R, P = V.I >

This formal model of the electric system may be implemented on computers for computing its behaviour. An important feature of formal models of physical systems is that human intentions do not play any role whatsoever in these models.

A formal representation of a technical artefact will have to take into account both its physical and functional features. I assume that the functional features may be treated formally in the same way as their physical features, namely as properties of (possessed by) the artefact (see also section 3.2). Consider the object of Fig. 5.7. It is a technical artefact, namely a heating device. It is a physical system with the technical function of heating things. If we take this object to be just a physical system (that is, forget about its technical function), and, more in particular, if we look at it from the point of view of the theory of electric circuits, then it is an object that is similar to the physical system of Fig. 5.6a. It is simply a resistance that may be connected to an electric power source. Thus, from that physical point of view the formal representation of this object will look similar to M(electric syst.).

Of course, it is possible to incorporate many more physical features into our formal model of the object of Fig. 5.7. We may include formal representations of its

[26] In order to get to a full-blown formal model a formal language will have to be constructed in which V1,..Vn and R1,…Rm are defined (which I will leave out here, since that is not necessary for my purposes).

Fig. 5.7 A picture of a technical artefact, namely a heating device. Considered as an electric circuit this object, when plugged in, is physically similar to the system of Fig. 5.6a

geometrical features, of the material features of its parts, of the laws governing the conductivity of heat in various substances etc. These formal models may be implemented on computers (for instance in CAD software) which makes it possible to simulate the physical behaviour of the object. However detailed these formal models of the physical features of the object may become, even if they include all physical features that are technically relevant, they do not capture its functional features. These physical simulation models may support a designer in studying whether under certain conditions (for a chosen set of values of design parameters) the physical behaviour of the technical artefact satisfies certain constraints (for instance, that the power generated by the resistance P is greater than or equal to a certain value P0). But the simulation model does not contain any information about the functional features of the technical artefact, for instance, that the function of the resistance is to transform electric power into heat, or that it ought to produce the amount of power P0. The simulation model describes the physical behaviour of the technical artefact (what it physically does) not what its function is (what it is for, or what it ought to do).

As a technical artefact the heating device of Fig. 5.7 also has functional features. These include[27]:

F(device):	for heating water (fluids)
H(resistance):	for transforming electric power into heat

[27] In order to distinguish them notationally from physical properties functional properties are denoted in bold.

S(surface): for supporting the object to be heated
M(surface): for conducting heat to the object to be heated, et cetera

Just as there are relations between the physical properties of an object, there are relations between the functional properties. For instance, the performance of the overall function of the object depends on the performance of the functions of its parts. One way to represent these functional relationships is with the help of a functional decomposition, which shows how the overall function of a technical artefact may be broken down into sub-functions of its parts. If we assume that the relationships between these functional properties may be expressed formally by the relations **R1**,…,**Rm**' we arrive at the following formal model of the heating device:

M(heating device) = M(electric syst.) + < **F, H, S, M…, R1,…,Rm'**>

As a formal model of a technical artefact this model still has a serious shortcoming. It represents the heating device as the sum of two independent sub-models, one for the physical and one for the functional features. It does not take into account that there are relations between the physical and functional features of the heating device. The function and physical structure of a technical artefact constrain each other. If we assume that these 'hybrid' relations between the physical properties V, I, R, P…and the functional properties **F**, **H**, **S**, **M**…may be formally represented by the relations *H1*,*H2*,..*Hj* we end up with the following formal model for the heating device:

M(heating device) = <V,I,R,P.., R1,..,Rn, **F,H,S,M.., R1,..,Rm**, *H1*,..*Hj*>

So, in general, a formal model of a technical artefact will have the following form:

M(technical artefact) = <P1,.. Pn, R1,..Rm, **F1,..,Fk**, **R1,..,Rl**, *H1*,..,*Hj*>

with P1,… Pn and R1,…,Rm referring respectively to physical properties and relations between these physical properties, **F1**,..,**Fk** and **R1**,..,**Rl** to functional properties and relations between these properties and *H1*,..,*Hj* to relations between physical and functional properties.

Against the background of these differences in formal models of physical objects and technical artefacts, various reasons may be proposed for explaining the difference in success in formally modelling both kinds of objects. One might try to explain the difference by claiming that there is a significant difference in *reliability* of our knowledge of physical and functional features. However, that may be questioned. Knowledge of functional features of technical artefacts, it seems, may be as objective as our knowledge of their physical features. For instance, the claim that this object on my desk is a screwdriver and has the function of fastening and loosening screws appears as objectively true as the claim that that same object has a certain mass (see the quote from Searle in section 2.5.2).[28] Just as its mass, its kind-proper function appears to be an objective feature of the object involved. Of course, not all statements about functional features are objective in this sense; it does not apply to ascriptions of use accidental functions.

In my opinion the reliability of functional knowledge is not the differentiating factor. What makes the two cases different is that the modelling of technical artefacts involves also the modelling of mind-dependent features, whereas the modelling of physical systems only involves the modelling of intrinsic, mind-independent features. Objects have functional features, including kind-proper functions, only by virtue of function assignments. This means that the formal representation of kind-proper functions requires a formal representation of kind-proper function assignments. It seems plausible to assume that such formalizations will have to make use of some formal model of beliefs, desires and intentions, often referred to as BDI-models; the development of these models, however, is still in its early stages. The situation with regard to physical features is different. These are intrinsic to the objects involved and therefore a formal representation of physical features does not require a formal treatment of some kind of assignment relation according to which physical features are assigned to technical artefacts or objects in general.

There is yet a second reason why it may be difficult to get formal modelling of technical artefacts off the ground. To see why, note that in formal models of technical artefacts the hybrid relations *H1,..,Hj* are of crucial importance because they establish a link between the physical and functional sub-models. In case formal representations of technical artefacts are intended to be used for supporting engineers in their functional reasoning about technical artefacts (which is one of the aims of formalizing functions mentioned earlier), these relations are of crucial importance. Among other things, this functional reasoning will have to help design engineers to explore the possible consequences of changes in physical properties for the functional properties (and vice versa). The nature of these hybrid relations, however, remains rather obscure. Several times we have made claims that imply relations between physical and functional features, such as the claim that the physical structure of a technical artefact has to realize or perform its function, or that function and structure cohere or constrain each other. Apart from the fact that a clear meaning of

[28] In Searle's terminology we are dealing here with epistemically objective judgments about ontologically subjective features; see Searle (1995).

these various relations is lacking, it is not clear at all whether, once clarified, a formal representation of these relations is within reach.[29]

5.8 The traditional design paradigm

In this final section I draw attention to what I call the traditional design paradigm and how it is put into question in fields of systems engineering where the boundaries of the kinds of systems they are designing, developing and implementing are extended into the direction of socio-technical systems (see section 2.1 and the Epilogue). Let me first take a closer look at the traditional design paradigm. It is based upon three assumptions about the kind of technical artefact that is designed. This kind is exemplified by stand alone consumer products. Many of the examples used so far fall into this category. These technical artefacts may be used by individuals or by groups, more or less in isolation of their wider technological and social context. What is required for the proper performance of their function is a technical artefact that does not malfunction and is properly implemented. To phrase it in Simon's terminology, the inner and outer environments of the technical artefact have to behave as they ought to. This brings me to the first important feature of the traditional design paradigm, namely the assumption that it is possible to clearly separate the object of design from its environment. In his analysis of engineering design Simon, for instance, simply assumes that this does not give rise to any problems. The second feature concerns an assumption about the nature of the object or system to be designed or, more to the point, the content of the inner environment. Traditional engineering concerns itself with the design of the hardware (the manual is more or less taken for granted).[30] What is designed is a material technical object. The final feature of the traditional design paradigm is that it is assumed that the behaviour of the technical artefacts designed can be fully controlled by controlling the behaviour of its parts, at least when the designed system is used under conditions specified within the design specifications. Given the second assumption, this control amounts to the control of the behaviour of physical parts (including their embedded software) through a set of control parameters. These three assumptions about the objects of design, which are not independent of each other, together characterise the traditional design paradigm.

Certain features of the kinds of systems designed within various fields of systems engineering undermine the applicability of the traditional design paradigm in these fields. Systems engineering arose in response to the ever more complex systems designed and developed by engineers (see, for instance, Sage (1992)). This development not only challenged engineers in relation to the designing of such complex systems but it also presented questions concerning the designing and organising of

[29] For an attempt to explicate the idea that there is a coherence relation between the structure and the function of a technical artefact, see Kroes (2006).

[30] Here I take hardware in the sense of also including embedded software.

the engineering-design process allied to such complex systems (Ottens et al. 2006). Here I concentrate on two features of the kinds of systems designed that pose questions in conjunction with the applicability of the traditional design paradigm. The first feature concerns the socio-technical nature of the systems designed, the second the possibility of emergent behaviour in complex systems.

One of the types of systems studied and designed within systems engineering are large-scale infrastructural systems, such as electric-power supply systems or public-transport systems. The behaviour of these systems is significantly affected by their technical elements but the functioning of the system as a whole depends as much on the functioning of these technical components as on the functioning of a social infrastructure (legal systems, billing systems, insurance systems etc.) and on the behaviour of human actors. From an engineering point of view this draws attention to the issue of whether the social infrastructure is to be regarded as part of the outer environment and modelled as a series of constraints for the design of technical systems or is to be taken as part of the system to be designed. An important argument in favour of including these social elements within the system is that the technological and social infrastructures have to be attuned to each other if such systems are to operate successfully. If social elements are included within the system, as is often advocated, then the implication is that systems engineering has to deal with the design and control of socio-technical systems. As we have seen in section 2.1, these are hybrid systems consisting of elements of various kinds, such as natural objects, technical artefacts, human actors and social entities like organisations and the rules and laws governing the behaviour of human actors and social entities.[31]

The traditional design paradigm no longer appears to be a suitable basic framework for the design and control of socio-technical systems. To begin with, there is the problem of where to draw the line between the system under consideration and its environment. This is a conceptual problem that systems engineering inherits from systems theory (Kroes et al. 2006). If the function of a system is taken to be that which gives the system cohesion, then it is rather obvious that all elements relevant to the functioning of the system should be included. So human agents and social institutions would have to become integral parts of the infrastructure systems alluded to above. But how is the function of, for instance, an electric-power supply system to be defined? Different actors may have different views on this and may therefore have different opinions on what constitutes part of the system and what belongs to its environment. The socio-technical nature of the systems designed also means that the nature of the system to be designed changes. The inner environment will no longer consist of only material objects. The design of these systems not only involves the design of technical but also of social infrastructures, including operator roles, to secure that they are tailor-made to match each other. Finally, the idea that these systems can be completely designed and controlled has to be abandoned. The behaviour of human agents and social institutions cannot be controlled in the

[31] Within the field of STS-studies these systems are often referred to as heterogeneous systems; see, for instance, Bijker et al. (1987).

way that the behaviour of technological systems can be controlled. In the traditional design paradigm it is assumed that there is a vantage point outside the designed system from which design and control is overseen. That is not the case with socio-technical systems in which various actors, with their own interpretations of the function of the system and their role in realising it, set out to change or re-design parts of the system from within. For this reason even the notion of the design of socio-technical systems becomes problematic.

The second feature of complex systems that threatens the applicability of the traditional design paradigm lies in the possible occurrence of emergent phenomena. In recent times, emergent phenomena in complex technological systems have become quite a topic of debate in engineering circles (Buchli and Santini 2005; Deguet et al. 2005; Johnson without date). The science and engineering of complex systems are turning into fields in their own right in which emergent phenomena are widely coming to be seen as a defining feature of complexity.[32] Complex systems may exhibit non-linear, chaotic behaviour that results in processes of self-organisation and in emergent systemic properties like adaptivity, robustness and self-repair (Bertuglia and Vaio 2005). From an engineering point of view such properties may be desirable but the drawback is that their occurrence may be unexpected and unpredictable. That makes it difficult to control such features. The desire to control emergent phenomena in complex systems is driven on the one hand by the fact that they may be dangerous (blackouts in electric power supply systems are often claimed to be such emergent features) and on the other hand by the fact that they may contribute to some desired property of complex technological systems (e.g. complex adaptive systems may be more robust in relation to changing conditions in the environment).[33]

Whether blackouts in electric-power supply systems are genuine examples of emergent phenomena or whether other real examples can be given, remains to be decided. Assuming, however, that emergent phenomena may occur in complex technical systems, they do pose a real challenge to the traditional design paradigm. This challenge is not related to the first and second features of this paradigm. Emergent behaviour may occur in systems where it is not problematic to establish where the boundary with the environment lies and where there is not necessarily evidence of 'hybrid' systems (although the socio-technical systems discussed above may prove to be a promising class of systems exhibiting emergent behaviour). It is the third feature, the assumption about the control of the behaviour of the system that has to be renounced with regard to emergent behaviour, if emergent behaviour is taken to be behaviour that cannot be reduced to the behaviour of the constituent parts of the

[32] See, for instance, the pre-proceedings of the Paris conference (14-18 November 2005) of the European Complex Systems Society, ECCS'05 (http://complexite.free.fr/ECCS/); this conference hosted satellite workshops on topics such as *Engineering with Complexity and Emergence* and *Embracing Complexity in Design*.

[33] Kasser and Palmer (2005) distinguish between three types of emergent properties namely undesired, serendipitous and desired; serendipitous features are described as "beneficial and desired once discovered but not part of the original specifications".

system. This means that techniques like functional decomposition cannot be applied to functional properties of systems that are based on emergent phenomena. It also implies that the behaviour of the system as a whole cannot be completely controlled by controlling the behaviour of its parts. So, emergence and control do not go hand in hand. According to Buchli and Santini (2005, p. 3) "there is a trade-off between self-organization [and emergence; P.K.] on one hand and specification or controllability on the other: if you increase the control over your system you will suppress self-organization capabilities." Such a new trade-off principle would indeed constitute a significant break with the traditional design paradigm.[34]

Given the growing complexity of the systems that engineers have to deal with, it is to be expected that systems engineering will become an ever more important branch of engineering. This growing complexity will pose new challenges to engineering-design practice. Whatever the precise nature of this complexity it will, without any doubt, stretch the applicability of traditional methods for designing and controlling technical systems to their limits or even beyond their limits. This means that for these systems the traditional design paradigm with its idea of 'total design control' may have to be left behind and alternative design paradigms may have to be developed instead.

5.9 Conclusion

The outcome of this chapter is partly a philosophical research agenda and partly a record of results obtained. This is mainly due to the fact that there is virtually no philosophy of making, in spite of all the recent efforts devoted to developing a philosophy of action. That is remarkable since making is a special form of action and not one of the least important, and accordingly one would expect it to be dealt with in the philosophy of action. That, unfortunately, is not the case and so the philosophy of making in general, and of making technical artefacts in particular, lacks a firm foundation. In my opinion, one of the greatest dangers for a philosophy of making technical artefacts is an intellectual bias, that is, to treat the making of technical artefacts primarily as an intellectual activity. Given the institutionalized division of labour within engineering practice between the mental and the physical aspects involved in making technical artefacts, between designing and producing them, and given, as I noted at the beginning of this chapter, a prevailing focus on design aspects, it is very alluring to treat the making of technical artefacts as primarily a mental activity. However, the making of technical artefacts is as much, or even more so, a physical activity, requiring physical work and getting your hands 'dirty', as it is a mental activity. Moreover, it remains to be seen whether from a philosophical point of view these two aspects of making can be really separated into consecutive phases. For a better insight into these issues, a philosophy of making is badly needed.

---

[34] For a more detailed analysis of the notion of emergence and its relation to the control issue in engineering, see Kroes (2009b).

What I have tried to do in this chapter is present an interpretation of engineering design that is in line with the dual-nature conception of technical artefacts as set out in the preceding chapters. Along the way I have pointed out a number of problems and ideas that need further philosophical analysis, among which the synthetic-analytic distinction referred to in the characterization of designing as a synthetic activity, as opposed to research as an analytical one, the analysis of means-end reasoning, that appears to play a crucial role in engineering design, and with regard to designing plans the question arose what rationality constraints are to be imposed on plans for making things. An examination of the notion of a design of a technical artefact showed that, from a philosophical as well as from an engineering point of view, this notion stands in need of further clarification. I have also put forward an explanation why the formal representation of technical artefacts, in particular their functional design features, turns out to be so difficult, in contrast to the formal representation of the physical properties of objects. Finally, I have discussed the limits of the traditional design paradigm, in particular in connection with the design of socio-technical systems.

References

Arendt, Hannah. 1958. *The human condition*. Chicago: The University of Chicago Press.

Barnes, Barry, David Bloor, and John Henry. 1996. *Scientific knowledge; a sociological analysis*. London: Athlone.

Beaney, Michael. 2007. Analysis. In *The Stanford Encyclopedia of Philosophy (Fall 2007 Edition)*, ed. Edward N. Zalta. (editor): http://plato.stanford.edu/archives/fall2007/entries/analysis/.

Bertuglia, Cristoforo Sergio, and Franco Vaio. 2005. *Nonlinearity, chaos, and complexity : the dynamics of natural and social systems*. Oxford ; New York: Oxford University Press.

Bijker, W.E., T.P. Hughes, and T. Pinch. 1987. The social construction of technological systems: new directions in the sociology and history of technology. Cambridge (Mass.): The MIT Press.

Bratman, Michael. 1987. *Intention, Plans, and Practical Reasoning*. Cambridge, MA: Harvard University Press.

Buchli, J., and C.C. Santini. 2005. *Complexity engineering: harnessing emergent phenomena as opportunities for engineering*. Reports of the Santa Fe Institute's Complex Systems Summer School 2005. Santa Fe: Santa Fe Institute.

Chittaro, Luca, and Amruth N. Kumar. 1998. Reasoning about function and its applications to engineering. *Artificial intelligence in engineering* 12:331–336.

Cross, Nigel. 1994 (1989). *Engineering design methods; strategies for product design*. second edition Aufl. Chichester: John Wiley & Sons.

De Ridder, Jeroen. 2007. *Reconstructing design, explaining artifacts; philosophical reflections on the design and explanation of technical artifacts*. Delft: Delft University of Technology.

Deguet, J., Y. Demazeau, and L. Magnin. 2005. Elements about the emergence issue; a survey of emergence definitions. In *Proceedings of the ECCS 2005*. Paris.

Dym, Clive L. 1994. *Engineering design; a synthesis of views*. Cambridge: Cambridge University Press.

Dym, Clive L., and Patrick Little. 2000. *Engineering design: a project-based introduction*. Chichester: John Wiley & Sons.

Eck, Dingmar van. 2011. Functional decompostion: on rationality and incommensurability in engineering. Delft University of Technology, Delft, Nl.

Fehér, M. 1993. The natural and the artificial. *Periodica Polytechnica; humanities and social sciences* 1 (1):67–76.

Franssen, M. 2005. Arrow's theorem, multi-criteria decision problems and multi-attribute design problems in engineering design. *Research in engineering design* 16:42–56.

Hacking, Ian. 1983. *Representing and intervening; introductory topics in the philosophy of natural science*. Cambridge: Cambridge University Press.

Houkes, Wybo. 2009. The nature of technological knowledge. In *Handbook of philosophy of technology and engineering sciences*, ed. Anthonie Meijers, 309–350. Amsterdam: Elsevier.

Houkes, Wybo N., Pieter E. Vermaas, Kees Dorst, and Marc J. De Vries. 2002. Design and use as plans: an action-theoretical account. *Design Studies* 23 (3):303–320.

Houkes, Wybo, and Pieter E. Vermaas. 2010. *Technical functions: on the use and design of artefacts*. Philosophy of engineering and technology. Dordrecht: Springer.

Hubka, Vladimir, and W. Ernst Eder. 1996. *Design science; introduction to the needs, scope, and organization of engineering design knowledge*. London: Springer.

Hughes, Jesse, Peter Kroes, and Sjoerd Zwart. 2007. A semantics for means-end relations. *Synthese* 158 (2):207–231.

Hughes, Jessy. 2009. Practical reasoning and engineering. In *Handbook of philosophy and engineering sciences*, ed. Anthonie Meijers, 375–402. Amsterdam: Elsevier.

Johnson, Christopher W. without date. What are emergent properties and how do they affect the engineering of complex systems? http://www.dcs.gla.ac.uk/~johnson/papers/emergence.pdf.

Kasser, J.E., and K.D. Palmer. 2005. Reducing and managing complexity by changing the boundaries of the system. In *Proceedings CSER 2005*. Hoboken, NJ, USA.

Kroes, Peter. 1996. Technical and contextual constraints in design; an essay on determinants of technological change. In *The role of design in the shaping of technology*, eds. J. Perrin, and D. Vinck, 43–76. COST A4, vol. 5; European research collaboration on the social shaping of technology.

Kroes, Peter. 2003. Physics, experiments and the concept of nature. In *The philosophy of scientific experimentation*, ed. H. Radder, 68–86. Pittsburgh: University of Pittsburgh Press.

Kroes, Peter. 2006. Coherence of structural and functional descriptions of technical artefacts. *Studies in History and Philosophy of Science* 37 (1):137–151.

Kroes, Peter. 2009a. Foundational issues of engineering design. In *Philosophy of technology and engineering sciences*, ed. Anthonie Meijers, 513–541. Elsevier.

Kroes, Peter. 2009b. Technical artifacts, engineering practice, and emergence. In *Functions in Biological and Artificial Worlds; Comparative Philosophical Perspectives*, eds. Ulrich Krohs, and Peter Kroes, 277–292. Cambridge, Mass.: MIT Press.

Kroes, Peter. 2010. Formalization of technical functions: why is that so difficult? In *TMCE (Ancona, Italy)*, eds. Imre Horvath, Ferruccio Madorli, and Soltan Rusak: Delft: Delft University of Technology.

Kroes, Peter, Maarten Franssen, and Louis Bucciarelli. 2009. Rationality in Design. In *Philosophy of Technology and Engineering Sciences*, ed. Anthonie Meijers, 565–600. Elsevier.

Kroes, Peter, Maarten Franssen, Ibo van de Poel, and Maarten Ottens. 2006. Treating socio-technical systems as engineering systems: some conceptual problems. *Systems research and behavioral science* 23:803–814.

Krohs, Ulrich. 2009. Functions as based on a concept of general design. *Synthese* 166 (1):69–89.

Mitcham, C. 1994. *Thinking through technology; the path between engineering and philosophy*. Chicago: The university of Chicago press.

Ottens, Maarten, Maarten Franssen, Peter Kroes, and Ibo van de Poel. 2006. Modelling infrastructures as socio-technical systems. *Int. J. Critical Infrastructures* 2 (2/3):133–145.

Pahl, G., and W. Beitz. 1996. *Engineering Design: A Systematic Approach*. London: Springer Verlag.

Paley, William. 2006 (1802). *Natural theology: or, evidence of the existence and attributes of the deity, collected from the appearances of nature*. Oxford: Eddy, M and Knight D.M. (eds.), Oxford University Press.

Petroski, Henry. 1992. *The Evolution of Useful Things*. New York: Alfred A. Knopf.

Pollock, John L. 2002. The logical foundations of means-end reasoning. In *Common Sense, Reasoning, and Rationality*, ed. Renee Elio. New York: Oxford University Press.

Quine, W.V. 1951. Two dogmas of empiricism. *The philosophical review* 60 (1):20–43.

Ratzsch, Del. 2005. Teleological arguments for God's existence. In *The Stanford Encyclopedia of Philosophy (Fall 2005 edition)*, ed. Edward N. Zalta. (editor): http://plato.stanford.edu/archives/fall2005/entries/teleological-arguments/.

Reymen, Isabelle. 2001. Improving design processes through structured reflection; a domain-independe approach. University of Technology Eindhoven; PhD Thesis, Eindhoven.

Rittel, Horst W.J., and Melvin M. Webber. 1984. Planning problems are wicked problems. In *Developments in design methodology*, ed. Nigel Cross, 135–144. Chichester: John Wiley & Sons.

Russell, Paul. 2005. Hume on Religion. In *The Stanford Encyclopedia of Philosophy (Fall 2005 edition)*, ed. Edward N. Zalta. (editor): http://plato.stanford.edu/arvhives/win2005/entries/hume-religion/.

Ryle, Gilbert. 1984. *The concept of mind*. University of Chicago Press Aufl. Chicago: University of Chicago Press.

Sage, A.P. 1992. *Systems engineering*. New York: John Wiley & Sons.

Searle, John. 1995. *The Construction of Social Reality*. London: Penguin Books.

Segerberg, Krister. 1992. Getting started: beginnings in the logic of action. *Studia Logica* 51: 347–378.

Simon, Herbert A. 1984. The structure of ill-structured problems. In *Developments in design methodology*, ed. Nigel Cross, 145–166. Chichester: John Wiley & Sons.

Simon, Herbert A. 1996 (1969). *The Sciences of the Artificial*. 3 rd ed. Aufl. Cambridge, MA: MIT Press.

Staudenmaier, John M. 1985. *Technology's storytellers; reweaving the human fabric*. Cambridge, Mass.: The MIT Press.

Stone, Robert B., and Kristin L. Wood. 2000. Development of a functional basis for design. *Journal of Mechanical Design* 122:359–370.

Thomasson, Amie L. 2007. Artifacts and human concepts. In *Creations of the mind: essays on artifacts and their representations*, eds. Stephen Laurence, and Eric Margolis, 52–73. Oxford: Oxford University Press.

Vermaas, P. E., and W. Houkes. 2006. Technical functions: a drawbridge between the intentional and structural natures of technical artefacts. *Studies in History and Philosophy of Science* 37 (1):5–18.

Vincenti, Walter G. . 1990. *What engineers know and how they know it*. Baltimore: John Hopkins U.P.

Von Wright, Georg Henrik. 1963. Practical inference. *The philosophical review* 72 (2):159–179.

Von Wright, Georg Henrik. 1972. On the so-called practical inference. *Acta Sociologica* 15:39–53.

Wood, William H. 2009. Computational representations of functions in engineering design. In *Handbook of philosophy of technology and engineering sciences*, ed. Anthonie Meijers, 543–564. Elsevier.

Quine, W.V. 1951. Two dogmas of empiricism. The philosophical review 60: 20–43.
[illegible]

Chapter 6
The moral significance of technical artefacts

"for there is nothing either good of bad, but thinking makes it so"

Hamlet, William Shakespeare

6.1 Introduction

The moral status of technical artefacts has been a matter of controversy for a long time. The main issue at stake is not how to assess technical artefacts morally but whether it makes sense at all to assess or evaluate them morally.[1] Human acts with or without technical artefacts may be assessed morally.[2] But what about the technical artefacts themselves? Are they susceptible to moral assessment? And if so, on account of what? It is quite common to assess technical artefacts in terms of their instrumental goodness and to make normative (evaluative) statements about them such as "This is a good knife". Here, however, I am not interested in this instrumental goodness of technical artefacts (see section 4.1), but in the possibilities of their moral goodness.[3] Does it make sense to call a technical artefact good or bad in a

[1] Of course, this issue involves the question what it means to have moral significance in general. This question is usually sidestepped in discussions about the moral significance of technical artefacts. I will do the same, but to give a rough indication, I will take whatever *action* that affects the well-being or flourishing of human beings in positive or negative ways to have moral significance. The question we are facing here is whether this idea of moral significance can be extended to *things*, in particular technical artefacts. The question about the moral significance of technical artefacts by themselves then boils down to the question whether technical artefacts by themselves affect the well-being or flourishing of human beings.

[2] Note that whether or not human acts, such as lying, are considered to have moral significance by themselves depends on the kind of ethical theory adopted. In utilitarian theories they have not (acts acquire moral significance through their consequences), whereas in deontological approaches they usually have (think of categorical imperatives such as "You ought not to lie").

[3] For a discussion of the varieties of goodness, see (Von Wright 1963b).

P. Kroes, *Technical Artefacts: Creations of Mind and Matter: A Philosophy of Engineering Design*, Philosophy of Engineering and Technology 6,
DOI 10.1007/978-94-007-3940-6_6,

moral sense? Only if this question is answered in a positive way does the question pop up of how to morally assess particular technical artefacts. As regards the latter question, it is not self-evident that traditional ethical theories with their focus on assessing human actions are relevant for assessing the moral significance of technical artefacts if indeed they do have such significance.

According to the well-known moral-neutrality thesis, technical artefacts are mere means for human beings to realize their ends, and any moral issue about technology is finally a moral issue about those ends and about human action. Technical artefacts by themselves have no moral significance, that is, play no role in moral deliberation or moral judgment about human ends and actions. Advocates of the neutrality thesis readily admit that technology constantly creates new options for action and makes it possible to realize ends that may be morally questionable. The moving front of state of the art technology constantly poses humankind for new moral problems. They maintain, however, that this does not affect the moral neutrality of the technical artefacts involved: once the new artefacts (options for action) are at hand, the moral discussion is about how they are to be used, about the ends that may be realized with the means, not about the means themselves.[4] They defend a strict separation of the domain of the means, which are to be evaluated in terms of instrumental goodness with the help of notions like efficacy and efficiency etc., from the domain of ends pursued by human beings, which are to be evaluated in terms of moral goodness.

This is a clear-cut view on the moral significance of technical artefacts, but apparently for many not very attractive since repeatedly people have argued against it.[5] One of the main arguments against this view is that it ignores that technical artefacts play such an important role in the life of human beings that they may be considered to be constitutive for being human. They shape human life and existence and determine the way people think and act to such a degree that it makes no longer sense to think of them as mere passive means, as a domain of things that may be separated from human beings. They *do* all kinds of things to human beings, also things that are morally relevant (Verbeek 2005). In Winner's terms, technology has become a 'form of life' (1986, Ch. 1). However, it is one thing to reject the neutrality thesis, it is another to put in place an alternative view on the moral significance of technical artefacts. In what sense may technical artefacts be said to actively shape and condition human existence such that they acquire moral significance themselves? Does the fact that technical artefacts are not mere passive means imply that it makes sense to apply to them concepts that have so far been confined to the description and interpretation of what humans do, of human action, such as the concepts of values, ends and moral agency?

---

[4] Whether the new options for action should have been created in the first place, raises itself moral questions that I will ignore here.

[5] It is not easy to find outspoken defenders of the neutrality thesis in the literature. Pitt (2000, forthcoming) is an example.

Over time, the issue of the moral significance of technical artefacts has been formulated in different ways. One of the traditional ways is to ask whether technical artefacts by themselves can be morally good or bad. Another is to ask whether they have inbuilt values or ends that are morally relevant. In recent times, the controversy has taken a new, rather radical turn by reformulating it in terms of moral agency. Does it make sense to attribute to technical artefacts some form of moral agency? The idea that technical artefacts are moral agents has gained considerable ground in STS- and post-phenomenological circles.

Clearly, these different ways of formulating the issue are related. When technical artefacts have inbuilt moral values or embody them, then they may be qualified as morally good or bad on the basis of these values. And when technical artefacts are taken to be moral agents then these agents may be assumed to have values and ends associated with them. I will not explore the connections between these different ways of formulating the issue of the moral status of technical artefacts.[6] They all appear to address a similar issue, namely whether moral goodness, moral values, moral ends or moral agency are to be located only in the context of design or use of technical artefacts or also in technical artefacts themselves. I refer to this issue as the issue of the moral significance of technical artefacts, and mainly focus on claims about technical artefacts having moral ends or moral agency by themselves. I defend the position that technical artefacts may have moral significance by themselves on account of the ends associated with their functions. These ends are inherent to technical artefacts and make them morally significant by themselves. They are, however, not intrinsic to technical artefacts because technical artefacts do not have ends independent of human agency; that is the reason for calling these ends *inherent* to technical artefacts. In this way it is, in my opinion, possible to account for the moral significance of technical artefacts without attributing moral agency to them.

6.2 Technological agency

There is a long-standing issue in the philosophy of technology that emerges time and again in different forms. It is the issue about how to interpret what technology *does* or technologies/technical artefacts *do* to people and society. The issue is about whether it makes sense to attribute to technology some form of agency independent of human agency. It is not controversial to claim that technology is some form of human agency or the outcome thereof. Pitt (2000), for instance, defines technology as a specific form of human agency, namely humanity at work. As a collection of technical artefacts, the origin of technology lies also undeniably in human agency; not only are technical artefacts produced by humans, but it is hard to make sense of the notion of a technical artefact without invoking the notion of human intentionality

[6] For instance, it is not clear at all whether claims about technical artefacts having moral agency by themselves imply claims about them having ends by themselves, or vice versa.

as I have argued in the first part of this book. What is controversial, however, is to claim that technology may have some form of agency of its own. Let me briefly point out some of the (at times metaphoric) ways in which the idea of technological agency has been put forward.

One form in which the idea of technological agency presents itself is in terms of master-slave tales about humankind and technology. The relation between humankind and technology may be conceptualized in a way that "man will be the ruling spirit and the machine the servant" (Butler 1985 (1872), p. 206). In this tale, technology belongs to the realm of means at the disposal of human agents for realizing their goals. In alternatives to this view, technology is depicted as a Golem, a human creation intended to serve human goals, which however turns against its creator and enslaves humankind. In Golem-like-tales technology sets its own laws (Ellul1964), becomes autonomous (Winner1992 (1977)), out of control, and starts to dominate humans, a domination "that is all the more perverse for not imposing the law of a master but that of an emancipated slave who does not have the least idea bout the moral goals proper to humankind" (Latour 2002, p. 247). It is out of fear for this domination by technology that the inhabitants of Samuel Butler's Erewhon have outlawed machines in their country.

Of course, the master-slave metaphor is not intended to be taken literally; technology does not act in the same way as human beings act. The metaphor is intended to convey the idea that technology escapes human control (human agency) and may influence human life in ways never intended by its creators or users. It is as if technology starts to live a life of its own, a life with a form of agency different from human agency, but some form of agency nevertheless.

The issue of technological agency is not confined to discussions about how technology as a whole affects human beings. In another form the same theme emerges in discussions that move at the level of technical artefacts. Winner's (1985) influential paper "Do artifacts have politics?" may serve as an illustration. Winner argues that technical artefacts may be *inherently* political in the sense that they require or are compatible with particular political structures. Though Winner does not claim that artefacts have agency, the title of his paper is telling. He defines politics as the (1985, p. 28) "arrangements of power and authority in human associations as well as the activities that take place within those associations." This notion makes sense within the domain of human agency, but is transplanted from that domain to the domain of technical artefacts. It is a metaphor, again, but the power of the metaphor consists precisely in the fact that it suggests that more is involved in the effects of technical artefacts on human beings than just human politics (or human intentions that are operative at the stage of design or use). Otherwise technical artefacts would have no inherent politics, no politics by themselves. Winner's thesis is interesting precisely because it posits that technical artefacts may have a form of politics that cannot be grounded in human intentions. Winner's metaphor suggests that, although he does not defend this claim himself, the politics of technical artefacts may have to be grounded in some form of agency of artefacts themselves.

In less metaphoric terms the idea of technical artefacts having some form of agency of their own is being discussed in the fields of ethics and technology and the field of STS. One of the issues at stake in these fields is the question whether moral issues about technology can be dealt with adequately by assuming that technical artefacts are just passive instruments for human beings to be used for the realization of their goals, and if not, whether it makes sense to attribute to technical artefacts a richer moral role by assuming that they have some form of agency. According to Huyke (2003, p. 57) the traditional discourse on technology, in which technology is treated as a collection of means or instruments, "tends to put the ethical burden exclusively on the human subject and not on technology x. The human subject may have questionable ends; technology x presumably does not even have ends". He claims that this conception of technologies obscures moral discussions and "that it is both cogent and ethically useful to describe them as agents in history, not as mere means" (2003, p. 58). Various proposals in that direction have been made. Ihde (1990), for instance, has put forward the suggestion to attribute a kind of intentionality to technical artefacts and Verbeek (2008) explicitly attributes some form of agency to technical artefacts. In the field of STS-studies the idea is widespread that technical artefacts are not just passive instruments but exhibit agency through scripts that are built into them (Akrich 1992; Latour 1992). Scripts in technical artefacts actively invite certain types of behaviour by their users. Within the context of these discussions it is often claimed that technical artefacts have morality or moral agency, either as objects on their own or in their relation to or combination with human beings.

Finally, there is the field of Artificial Intelligence (AI), which aims to build agency into technical artefacts. Within the AI-field one of the main problems is whether complex, computer-based systems may show behaviour that is comparable to human behaviour and therefore may be considered to be 'artificial' agents. To deal with this problem it is necessary to analyse what kind of behaviour is specific for human beings. What is needed is a set of conditions each of which is necessary and together sufficient for human agency (see e.g. Floridi and Sanders (2004)). Such discussions are largely absent when agency is attributed to technical artefacts, both in the field of ethics and technology and in STS. This is not really surprising. The basic problem of the moral/political significance of technical artefacts in the latter fields is not confined to a particular type of technical artefact (complex computer-based systems), but to technical artefacts in general. The examples used to illustrate agency often refer to relatively simple technical artefacts like street bumps, seat belts, guns and turn pikes. These do not, at least at first sight, show behaviour that is comparable to behaviour associated with human agency (it seems quite a challenge to argue that the behaviour of such artefacts somehow springs from beliefs, desires and intentions they have or is some form of autonomous action[7]) and are therefore not of prime interest for the issue of artificial agents. In the field of AI the focus is on building human agency, at least in its behavioural manifestation, into complex (computer) systems.[8]

[7] Verbeek (2008) argues that it makes sense to attribute a form of autonomy and freedom to technical artefacts.

[8] For a discussion of the possibilities of constructing moral artificial agents, see for instance (Wiegel 2007).

This rather impressionistic and fragmentary survey illustrates that different concerns and problems lie at the root of the idea of technological agency. With regard to technology as a whole, the main concern appears to be the unpredictability and uncontrollability of the effects of (often massive) use of technologies. Technology has all kinds of *unintended* effects, that is, unintended by human beings. These effects are then attributed to the agency and intentions of the Golem of autonomous technology. At the bottom of the discussion about whether technical artefacts have politics lies another concern. Here not the unintended effects of technical artefacts are at stake. According to Winner's version of the famous 'bridges of Long Island' example, their designer Robert Moses explicitly intended the effects. Winner's concern is with the fact that Moses's politics (intentions) are frozen into the bridges and that people may be exposed to Moses's politics for "generations after Moses has gone and the alliances he forged have fallen apart" (Winner 1985, p. 28). Another concern of Winner is that certain types of technology, e.g., nuclear power, as a matter of practical necessity require a certain shaping of the social order and in that sense are inherently political (which is not the case with regard to the bridges, for there are no technical reasons why these could not have been designed with a more egalitarian politics in mind). The discussion about artefacts as moral agents is mainly driven by the concern how to make sense of the moral significance of technical artefacts. Many philosophers of technology have problems with the idea that technology is simply a collection of morally neutral instruments to be used for morally good or bad purposes. They claim that the moral significance of technology goes much deeper. Just as politics, morality may be "frozen", "inscribed" or "embodied" into technical artefacts, and this makes them in one sense or another moral agents.

Although various concerns drive these debates, there is one problem that seems to be involved always (apart from the AI-case), namely: How to interpret the influence of technology on human beings (individually or collectively)? How to interpret what technology *does*, or what artefacts *do*? Schematically, three strategies for dealing with this issue may be distinguished.

1. The first one is that technology does nothing of itself and that any alleged agency of technology can be traced back and reduced to human agency. Technology itself is not active in any way independent of human activity. It belongs to the realm of passive means. This does not mean that technical artefacts may not causally influence the behaviour of human beings. They do, of course, just as natural objects do, but they do not cause anything by themselves, as in the case of human agency; they are not autonomous agents that take decisions that give rise to causal chains *originating* from these artefacts themselves. This is the point of view underlying ideas of the moral neutrality of technology and theories of social determination of technology.

2. According to the second strategy technology exhibits agency in the sense that it causes things, sets things in motion in ways that cannot be traced back to human agency. The effect of technology on human beings is more than simply a reflection of human activity onto itself; technology itself is a source of activity and not just a medium for propagating human activity. This technological agency is not necessarily of the same kind as human agency. If it is assumed that human agency is powerless

in the face of technological agency, then this strategy may easily lead to the idea of technological determinism or autonomous technology.

3. The third strategy is of a rather radical kind in the sense that it undermines or even rejects the way the question has been posed. It shifts the primary locus of agency from human beings and/or technology separately to human-technology associations or humans and non-humans in collectives or actor-networks (Latour 2005; Verbeek 2008). In these associations or networks humans and technical artefacts "co-constitute" each other. Neither humans nor technical artefact exist and act by themselves and therefore the original question is posed in a wrong way.

These different views on technological agency in general influence the debate about the moral status of technical artefacts. Roughly the positions taken in this debate run parallel to these three strategies. In the following sections I present and analyse these positions in more detail and argue that each shows serious shortcomings. Thereafter I present and alternative approach that does not fit into one of the above strategies.

6.3 Technical artefacts without intrinsic moral significance

"Guns don't kill people, people kill people", so the slogan of the American National Riffle Association (NRA) runs. Guns don't act and whatever atrocities are committed with guns, the human beings involved are to blame or responsible, for they and only they act. Guns may be used in ways that are morally good or bad, but nothing in guns themselves makes them morally good or bad things. Moral issues about guns are to be located and settled in their context of use. It is a category mistake to call guns morally good or bad, just as it would be a category mistake to call electrons morally good or bad. They are simply morally neutral, passive instruments.

Before I proceed to an analysis of the main premises on which this position is based, let me clarify once more what is at stake. The issue is not whether guns may be used for morally good or bad ends. The question is whether guns as such may be considered to be morally good or bad. Let us assume for the sake of argument that guns as such have moral significance, for instance, that they, as guns, may be qualified as morally bad technical artefacts. Then in spite of this it is not excluded that a gun, as a gun, may be used for morally good purposes. This is more or less analogous to the situation with regard to the moral significance of human acts. Lying as such may be considered to be morally bad, but that does not exclude the possibility that in a specific context or situation committing an act of lying may be morally good. What is at issue here is whether, in analogy to human actions, technical artefacts by themselves may have moral significance.

Let me outline what I take to be the two main premises on which arguments for the morally neutral status of technical artefacts, such as the above one, are usually based. These premises stay often implicit but nevertheless appear to play an important role in the neutrality quarters. I will formulate these premises in a rather stringent way, which makes me vulnerable to the objection of constructing a straw man.

However this may be, unearthing these two premises elucidates what is principally at stake in the debate from the point of view of defenders of the neutrality thesis.

The first premise concerns the strict separation of the ends that may be achieved with the help of technical artefacts from those artefacts. The ends are always the ends of human beings using technical artefacts and may, according to this position, be evaluated sensibly from a moral point of view. In contrast, technical artefacts themselves, as means to certain ends, may be evaluated from the point of view of practical rationality (are they effective/efficient means?), but not morally. They have no ends in themselves and therefore it simply does not make sense, to evaluate them from a moral point of view, just as it does not make sense to morally evaluate physical objects.

A second, related premise states that technical artefacts are just physical objects, albeit (in most cases) physical objects with a special history, namely they are human-made physical objects or constructions. From a moral point of view, however, this history does not set them apart from other physical objects. Technical artefacts therefore have the same moral status as physical objects in general (such as an electron). The physical properties (or causal capacities) of these human-made physical constructions are exploited to realize certain ends. According to this point of view, the claim that a technical artefact has a certain function amounts to the claim that it has (or may be expected to have) the physical capacity that enables users to realize the end associated with that function. However these ends may be evaluated morally, these functional properties, i.e., physical capacities, by themselves have no moral significance.

Arguments for the moral neutrality of technical artefacts along these lines are, in my opinion, seriously flawed because they are based on inadequate notions of a technical function and of a technical artefact (kind). The main problem with this view is that it identifies functions of technical artefacts with their physical capacities. As we have seen in chapters 2 to 4, however, this raises serious problems about interpreting malfunction of technical artefacts and being an instance of a technical artefact kind. The interpretation of technical artefacts as human-made-physical-objects is unable to account for the idea that technical artefacts come in kinds. It is precisely this shortcoming that plays a crucial role in the argument for the morally neutral status of technical artefacts.

Suppose that a gun is used successfully to hammer a nail in a wall. This use does not turn the physical object *X*, corresponding to the gun, into a technical artefact of the kind 'hammer'. It makes sense to claim that *X* is used *as* a hammer, but as we have seen before being used as a hammer is not a sufficient (or necessary) condition for *being* a hammer. In order to account for *X* being a gun and not a hammer, an appeal has to be made to the distinction between the kind-proper function of *X* and its use-accidental function (see chapter 4). But how is this distinction between kind-proper function and use-accidental function to be made? On this point the human-made-physical-construction conception of technical artefacts runs into difficulties. The answer cannot lie within the causal capacities (physical structure) of the object *X* involved. *X* has generally speaking many causal capacities that may be exploited for realizing various ends (among which the physical capacity to fire bullets but also the capacity to hammer with), and none of these causal capacities of *X*, considered as a physical object, has a preferred status. So the distinction between the kind-proper function of *X* and its use-accidental functions cannot be grounded in its causal capacities; these are all on an equal footing. From the point of view of its physical capacities there is nothing in the object *X* itself that makes it a gun rather than a hammer.

It is this consequence of the human-made-physical-object conception of technical artefacts that guarantees that they have a morally neutral status. None of all the possible functions that may be attributed to a particular human-made physical construction has a preferred status, because none of its physical capacities has. This means that the object *X* is no more a gun than a hammer and therefore there can be no special relationship between that object and the end associated with one of its particular functions (physical capacities). The context of use determines which one of all possible functions that a physical construction may fulfil is actually chosen, and so also which one of the ends associated with all possible functions is being realized. This is the reason why the idea of a moral evaluation of technical artefacts by themselves makes no sense. Technical artefacts as human-made-physical-constructions have by themselves no ends associated with them, on the basis of which they might be morally assessed.

6.4 Technical artefacts with intrinsic moral significance

"Guns kill people" advocates of gun control claim against the NRA. Clearly, these advocates do not intend to say that guns by themselves shoot people without somebody pulling the trigger. What they are claiming is that guns may *make* people kill each other, because people behave differently with or without a gun. This change in behaviour is not only due to the physical/functional characteristics of a gun but also, and may be primarily, to the *meaning* attached to the gun. A gun is not only an instrument of physical power, but also a symbol of power. As a consequence, a gun may 'transform' its owner into a murderer: a quarrel that without a gun would have been settled by a fist fight now ends in a killing. So, guns do things to people that go far beyond them being mere passive instruments for human beings to realize their goals and this makes that they have moral significance by themselves.

One way to interpret this kind of reasoning is to assume that technical artefacts play a genuine *active* role and that their activity cannot be interpreted solely in terms of human agency. It is this activity that brings with it a moral significance of technical artefacts on their own. In order to account for this moral significance a concept of technical artefact is needed that is different from a human-made physical object. A technical artefact must have properties over and beyond it physical capacities in order to make sense of its active role and moral status. One candidate for such a property may be the function of a technical artefact, if the latter is interpreted as a non-physical property.[9] Technical artefacts, it may be argued, have functional properties of their own that are as real as their physical properties; that a certain object is a gun is as much a real fact about that object as that it weighs one kilogram, and yet it cannot be reduced to its physical properties. As we will see later on in more detail (in section 6.7), technical functions may be intimately tied to symbolic functions, more in particular to meanings. So, technical functions often come with meanings attached to them. This meaning may be yet another candidate for a non-physical property of technical artefacts that may bestow moral significance on them. Besides function (and meaning), other non-physical properties have been considered in the literature as the basis for the moral significance of technical artefacts by themselves, such as technological intentionality (Ihde 1990; Johnson 2006) or inbuilt scripts (Akrich1992; Latour1992). These non-physical properties may form the basis for justifying the idea that technical artefacts have moral significance of their own, quite apart from any moral issues related to their actual contexts of use. If a technical artefact has a function or an inbuilt script, then that artefact is intimately tied to specific ends associated with that function or script and through those ends the technical artefact may acquire moral significance on its own.

The main problem with such proposals concerns a clarification of the nature of these non-physical properties. What kinds of non-physical properties are functions, meanings, technological intentionality or scripts, how do they relate to the physical ones, and to what extent are these non-physical properties independent of human agency? More in particular, a clarification of these notions is needed in relation to the idea that technical artefacts may have some genuine form of agency. For instance, in what sense can a technical artefact through its inbuilt script 'invite' or 'enforce' users to certain types of behaviour and what model of agency lies at the bottom of such claims?

Let us restrict ourselves to technical functions and briefly explore whether it is possible to argue for a moral significance of technical artefacts by themselves on the basis of their (non-physical) functional properties. If technical artefacts are supposed to have moral significance by themselves, by virtue of their functional properties, then these functional properties have to be properties artefacts have by themselves. The crucial question is how the expression "by themselves" is to be interpreted. Since I am interested in analysing the idea that technical artefacts have moral significance (agency) independently of human agency, the most obvious interpretation of that expression is: independently of human intentions and human agency. In other words, in order to sustain the argument of technical artefacts having moral significance (agency) of their own, functional properties have to be taken as intrinsic, mind-independent properties of technical artefacts. That leads to an

---

[9] I do not pretend that conceptions of technical artefacts, in the form presented here, have been explicitly defended in the literature, although I think they are lurking in the background of many arguments against the neutrality thesis. Again my aim is to bring to the fore as clearly as possible assumptions about technical artefacts underlying the idea that they have moral significance by themselves.

interpretation of the notion of a technical artefact according to which an object may be a technical artefact independently of human intentionality or agency. I have argued extensively against such a concept of technical artefacts; it does not take due account of their dual nature. The concept of technical artefact upon which this line of reasoning for intrinsic moral significance is based appears seriously defective.

Reviewing the discussion so far, I conclude that both sides in the debate on the moral significance of technical artefacts employ inadequate concepts of technical artefacts. On one side, they are treated as human-made physical objects that have from a moral point of view the same status as physical objects in general, on the other, technical artefacts are physical objects endowed with non-physical properties that form the basis for their moral agency or value. The first identifies technical functions with intrinsic physical properties (capacities), the second with intrinsic non-physical properties. Both approaches have in common that they assume that it makes sense to consider technical artefacts by themselves, in the sense of objects isolated from intentional human action. What makes an object a technical artefact are its intrinsic, mind-independent properties. That is also their common mistake. According to the-dual nature conception of technical artefacts, they are mind-dependent entities by definition. This means that the question about the moral significance of technical artefacts by themselves is put in a wrong way. Before I analyse the moral significance of technical artefacts from their dual-nature point of view, I discuss Latour's analysis of the moral status of technical artefacts. He too claims that technical artefacts cannot be considered in isolation from human agency, but for a different reason, namely because technical artefacts and human beings co-constitute each other.

6.5 Technical artefacts as moral agents

The question we started off with is framed in a wrong way when it is assumed that technical artefacts by themselves and human beings by themselves do not exist because they co-constitute each other.[10] In that case, the starting point for analysing the moral significance of technical artefacts cannot be the isolated technical artefact by itself. In recent times, several proposals have been put forward to shift the locus of analysis from the isolated technical artefact to collectives, associations or systems of humans and technical artefacts (Latour 1992; Achterhuis 1995; Latour 2002, 2005; Verbeek 2005, 2008; Introna forthcoming). These proposals have in common the idea that the moral significance of technical artefacts shows itself primarily at the level of these wholes. Moral agency is considered to be distributed, one way or another, over the elements of these wholes, elements that co-constitute each other by being part of these wholes. Here I concentrate on the work of Latour because it has proven to be the most influential.

[10] See, for instance, Latour (2002, p. 156): "Nothing, not even the human, is for itself or by itself, but always *by other things* and *for other things*."

Let me first sketch the broader framework, called the actor-network-theory, within which Latour analyses the morality of technical artefacts (Latour 1993, 1999, 2005).[11] His starting point is to give up the modernist subject-object distinction and to put the distinction between humans and non-humans (to which natural objects and technical artefacts belong) in its place. He treats technical artefacts (non-humans) and humans symmetrically, which means that he does not want to impose on them a priori distinctions, such as the distinction between intentional human action and causal action. Technical artefacts and humans are nodes in networks of relations. Neither technical artefacts nor human beings enter as pre-given entities in these networks. The essences of these entities are constructed in stabilizing networks; they are the output of these networks, not their input. Technical artefacts and human beings co-constitute each other in these networks (also called collectives or associations of humans and non-humans). In keeping with his symmetrical approach, he claims that both technical artefacts and humans show agency. Just as human beings may act morally, also technical artefacts may act morally and there is no great divide with technical artefacts on the side of the means, and the human on the side of the ends (Latour 2002, p. 248):

> In any case, the image of a human being at the helm manipulating inert objects to achieve ends through the intermediary of 'efficient action on matter' appears increasingly muddled. Technologies belong to the human world in a modality other than that of instrumentality, efficiency or materiality. A being that was artificially torn away from such a dwelling, from this technical cradle, could in no way be a moral being, since it would have ceased to be human – and, besides, it would for a long time have ceased to exist. Technologies and moralities happen to be indissolubly mingled because, in both cases, the question of the relation of ends and means is profoundly problematized.

Technical artefacts may be as much means for human ends, as humans may be means for the ends of technical artefacts. Agency, also moral agency, is distributed over all actors in a network; properly speaking only associated entities act (Latour 1999, p. 182).

Anything that makes a difference in the way other actors behave is an actor. Actors can make other actors do things; in particular they can delegate actions to other actors. But these other actors may overtake action and may behave in unexpected ways. Here one of Latour's main concerns comes to the fore, namely, how to come to grips with the uncertainty surrounding action. In *Reassembling the social* (2005) he distinguishes five sources of uncertainty. The uncertainty about the outcome of putting technical artefacts into the world is but one of these forms of uncertainty. The way he deals with this uncertainty is by letting technical artefacts themselves become things that act by making other actors do unexpected things. That is why technical artefacts are mediators and not simply intermediaries (Latour 2005, p. 39). Intermediaries transport meaning and force without altering them; given their input, their output may be predicted. That is not the case for mediators: their input is not a good predictor for their output; they "transform, translate, distort, and modify the

[11] For a summary of Latour's work, see Verbeek (2005, Chapter 5).

meaning or the elements they are supposed to carry" (*ibidem*). Technical artefacts as mediators are therefore a source of uncertainty and surprise. This brief sketch hardly does justice to Latour's general framework, but for our purposes it will do.

The moral agency of technical artefacts is worked out in more detail as follows. Humans may delegate parts of their programs of actions to technical artefacts (the reverse is also true). Latour's example of a speed bump illustrates how this works. There are several ways in which a municipality may enforce car drivers to drive slowly at a dangerous spot on a road.[12] One way to see to it that they do so is by putting a policeman at that spot, or someone waving a red flag. Then the program of action of slowing down traffic at a certain point is delegated to human beings. Another way is the put up a sign for a speed limit. It is also possible, however, to delegate this task to a technical artefact, namely a speed bump. The program of action is then inscribed into matter. In this way, the program of action is shifted from a human to a non-human, namely a technical artefact (Latour 1992, p. 244). The end result is that "stubborn and efficient machines and mechanisms" discipline the behaviour of car drivers. A technical artefact like a speed bump has an inbuilt script that prescribes drivers to drive slowly. They enforce the desired behaviour by a very effective program of action: if a driver does not follow the prescription (s)he ends up with a wrecked car suspension.

Thus technical artefacts, like human beings, may execute programs of action and are therefore agents, just like human beings. Being agents similar to human agents they may also act morally. For Latour the moral action of technical artefacts is related to the prescriptions inscribed in them (Latour 1992, p. 232):

> I will call [...] the behaviour imposed back onto the human by nonhuman delegates *prescription*. Prescription is the moral and ethical dimension of mechanisms. [...] We have been able to delegate to nonhumans not only force as we have known it for centuries but also values, duties, and ethics.

It is, for instance, the moral duty of the speed bump, in its association with car drivers and road authorities and other actors involved, to slow down traffic. In associations or assemblies of humans and technical artefacts the latter are to be seen as full-fledged moral agents. Morality is as much to be found in technical artefacts as in humans. Technical artefacts are what Latour calls the missing masses of sociology. Social relations and moral rules are not sufficient to account for what binds us together and makes us behave properly. To account for society, sociologists have to turn to non-humans, to technical artefacts. They are the "hidden and despised social masses who make up our morality" (1992, p. 227).

It is not my intention to take issue here with Latour's broader framework as such, in particular with his symmetric treatment of technical artefacts and humans. I am primarily interested in a critical assessment of his analysis of the morality of technical artefacts; that, however, will give us cause to question some aspects of his broader framework.

---

[12] See also the discussion about social, technical and socio-technical solutions to practical problems in section 2.1.

The first point to note is that there is something deeply disturbing about Latour's concept of morality of technical artefacts. This morality is rooted in the prescriptions built in to them. However, what is so specifically moral about them? Why is the prescription to slow down at a speed bump to be qualified as moral? Not all prescriptions are moral. In order to be able to listen to music on a CD I have to follow the prescriptions for using my CD-player. Why are these prescriptions moral? Why is the prescription of Latour's desk that in order to open one drawer the other two have to be closed a moral prescription (Latour 2002, p. 253).[13] Latour complains that he is 'obliged' to obey this moral law, but it is not clear at all what is moral about this law.

The prescriptions of technical artefacts may be interpreted as practical necessities. Practical inferences allow as to derive these practical necessities from statements of an end and of causal relationships (see our analysis of means-end reasoning in section 5.4; see also Von Wright (1963a, 1963b)). For instance, the prescription to slow down for a speed bump may be derived from the following two statements:

The driver does not want to damage the suspension of his car.
Unless the driver will lower his speed before crossing the speed bump, the <u>suspension of his car will be damaged.</u>
Therefore the driver must slow down before crossing the speed bump.

Note that the causal relationship in this example is valid for the driver's car. There may be cars for which this causal relationship is not valid; consequently, there would be no practical necessity for the driver to slow down for the speed bump. Or drivers may consciously develop what Latour calls anti-programs, by fitting the car with an extra strong suspension, which makes it possible not to subscribe to the prescription of the speed bump, i.e., to drive over speed bumps at high speeds. In those cases, the speed bump cannot enforce the required behaviour simply by its physical characteristics. For those cars, the practical inference is no longer valid and there is no practical necessity for drivers to slow down (and so other means to enforce slowing down will have to be put in place).

The 'must' in the prescription is not to be confused with a moral ought, with a moral obligation. According to Von Wright the conclusion describes a practical necessity. In designing and making technical artefacts engineers constantly have to deal with prescriptions of this kind. These prescriptions are, however, not moral simply because they are prescriptions. Latour's complaint about the fact that he is "'obliged' to obey the moral law of his desk, may also be taken to be a complaint about a practical necessity. Of course, this practical necessity is contingent upon the design of the desk. It could have been designed in another way, such that it would

[13] A similar conflation of practical necessity with the moral is to be found in Winner (1985, p. 34). He speaks of "the moral claims of practical necessity" and "moral reasons other than those of practical necessity".

not be necessary to have the other two drawers closed before opening the third one. That, however, in no way changes the nature of the practical necessity about opening and closing drawers of Latour's desk; that does not make it into a moral law. If the design of the desk is the object of Latour's moral complaint, then it is to be addressed to the designer, not to the 'moral law" of the desk.

The problem with regard to Latour's use of the notion of morality of artefacts is compounded by problems about his notions of action, actor and agency. It is to be expected that these notions will be different from the ones used in traditional ethical accounts, since he enlarges the domain of moral agents.[14] His symmetry principle requires that we assume that humans and technical artefacts in their associations act alike; both act in moral ways. But what does it mean for technical artefacts to *act* over and above *causal interaction* with their environment? In what sense can they be said to act in symbolic ways (through meanings)? How is human action to be interpreted from a symmetric point of view? It seems that it is not possible to hang on to the intentional aspect of human action, for that would imply attributing goals and intentions to technical artefacts (for instance, a gun).[15] Apart from these problems there is the distributed character of agency and the fact that there is always uncertainty about the origin of action, about who or what is acting in a given association (Latour 2005, p. 46). In his analysis of someone shooting with a gun, Latour remarks that it is the hybrid consisting of gun and gunman that acts. He claims that neither guns nor people kill and that the responsibility for the shooting is to be located in all the actors involved.

One of the most problematic aspects of Latour's view on technical artefacts and their morality concerns the identity of technical artefacts across different associations. Latour wants to give technical artefacts a more dignified place in the ontology in the world by making them moral agents (Latour 2002, p. 257); technical artefacts, he maintains, deserve better than to be treated as things (Latour1999, p. 214). However, not only is it unclear what the gift consist of, but the gift also comes with

[14] For a discussion of various notions of action used in moral theory, see (Schapiro 2001). Johnson (2006, p. 198) discusses five conditions human behaviour must meet in order to count as action and to be susceptible to moral assessment.

[15] Latour is not clear on this point. For instance, he claims that both humans and non-humans can have goals (which he seems to equate with functions) (1999, p. 180) but at the same time denies that purposeful action and intentionality are properties of human beings (1999, p. 192). See also the discussion in Pickering (1995, p. 17–18); Pickering endorses many aspects of actor-network-theory but parts ways with it when it comes to applying the symmetry principle to agency of humans and non-humans. He refuses to accept intentionality in the case of agency of non-humans. Nevertheless he insists on speaking about agency of non-humans. He considers, for instance, a lathe to be (1995, p. 158) "a prototypical device for capturing nonhuman agency: one can accomplish things with a lathe that naked human agency could never accomplish." I have problems with even such a limited notion of agency. Instead of a lathe, take a lever or a hammer. All that Pickering says about the lathe can also be said about a lever or hammer. Do they have nonhuman agency? I cannot think of any more passive technical artefacts than levers or hammers. They do nothing of, by or in themselves.

a considerable price, because technical artefacts (and human beings also) have to give up their identity (Latour1999, p. 179–180)[16]:

> You are different with a gun in your hand; the gun is different with you holding it. You are another subject because you hold the gun; the gun is another object because it has entered into a relationship with you. The gun is no longer the gun-in-the-armory or the gun-in-the-drawer or the gun-in-the-pocket, but the gun-in-your-hand, aimed at someone who is screaming. What is true of the subject, of the gunman, is as true of the object, of the gun that is held.

On such an account, any prosecutor will have great difficulty in proving that the gun-on-the-desk-in-the-courthouse is the same as the gun-with-which-Kennedy-was-murdered or that the suspect-in-the-courthouse is the person-who-shot-the-gun. That is problematic, to say the least. Whatever conceptual framework for the relation between humans and technical artefacts is proposed, it must be able to deal, in my opinion, with the possibility to establish the identity of objects and humans moving from one collective to another.[17] Otherwise it will be difficult to makes sense of moral aspects of everyday practices. To assume such an identity across associations, however, goes against the basic assumption that humans and technical artefacts co-constitute each other in collectives or associations.

Clearly, from Latour's perspective the problem we began with is framed in the wrong way. That problem is how to interpret the moral status of technical artefacts by themselves; technical artefacts, once put into the world, appear to start a life of their own, which means that they will 'do' unexpected things. Latour deals with this uncertainty by attributing agency to technical artefacts in their associations with human beings. On traditional accounts of agency in ethics this means that technical artefacts, in their associations with human beings, may be the sources of new causal chains or of new meaningful assertions just as human beings may be.[18] Such an interpretation of moral agency of technical artefacts is difficult to defend. Apparently, this is not what Latour intends to say since he rejects the traditional accounts of agency in the context of human beings using technical artefacts. Only associations act in a genuine sense and this action is distributed over the elements in the association. Such an account of (moral) agency does not bring us any closer to a better understanding of the moral significance of technical artefacts. I agree with Latour (2005, p. 51) that agency is about one of the most difficult problems philosophers have to deal with and a brief look at the philosophy of mind and of action suffices to see that traditional accounts of agency, starting off from the distinction between subject and object, may get one into deep trouble when dealing with some of the issues raised by Latour. Maybe some of these problems may ultimately force us to give up the distinction between subject and object. Be that as it may, I propose an

[16] For an analysis of the gun example similar to Latour's, see Verbeek (2008).

[17] Lurking behind the scenes, here, is of course the problem of the distinction between essential and contingent properties of things. I assume that Latour would also reject this distinction because it is an *a priori* distinction and essences are constructed in associations or collectives.

[18] For the interpretation of actions as meaningful assertions, see Schapiro (2001).

approach to deal with the moral status of technical artefacts that avoids making them actors like human beings.[19]

Let me finish with a remark about the claim that human beings and technical artefacts co-constitute each other, a claim echoed by many other people. Even if it is admitted that technology *in general* is a defining feature of humankind, it is, in my opinion, not necessary to conclude that technical artefacts and human beings co-constitute each other in each specific case of a human being using a technical artefact. I agree with Latour, and others, that technology in general is constitutive for modern human beings; it is hard, if not impossible, to think of humankind without technology; being embedded in technology may rightfully be claimed to be an essential feature of being human. But that does not mean that we have to draw the conclusion that at the level of individual technical artefacts human beings and technical artefacts co-constitute each other: that the person with the gun becomes a genuinely different person. To draw such a conclusion is to mix up different levels of analysis, to confuse the part with the whole.

6.6 Technical artefacts with inherent moral significance

For Latour, the proliferation of hybrids, objects that are neither human (social/intentional) nor non-human (material/natural) is a sure sign that the distinction between humans and non-humans has long passed its expiration date. In my attempt to interpret the moral significance of technical artefacts I precisely intend to take the hybrid character of technical artefacts seriously. This means that I will not put into question the great divide between intentionally acting human beings on the one side and causally interacting material/physical objects on the other. Technical artefacts, as I have argued extensively before are hybrid objects that are neither at home in the world of material objects, nor in the world of intentionally acting agents. Their proper place is in both of these two worlds at the same time. The failure of the attempts to make sense of their moral significance by treating them either as passive material objects (see sect. 6.3), or objects with intrinsic non-physical (intentional) properties (see sect. 6.4) is due partly to the fact that these attempts try to force technical artefacts in either of these two worlds in which they do not fit.

From the point of view of the dual nature of technical artefacts, the issue of their moral significance has to be put in a new perspective. It makes no sense to consider technical artefacts on their own, that is, as objects isolated from human agency or human intentionality. Doing so either leads to the idea that technical artefacts are just physical objects, for which the moral-neutrality thesis follows immediately, or to the idea that technical artefacts have moral significance by virtue of some allegedly intrinsic properties, such as their function or scripts. The function of a technical

[19] For a discussion of other kinds of objections against attributing morality to technical artefacts, see (Verbeek 2005, p. 213 ff).

artefact, however, is not an intrinsic property; it is a property that only makes sense in relation to human intentions. Nevertheless, in so far as the function is taken to be a defining feature of being a particular artefact (that is, without its function what is left of the artefact is just some physical object or construction), this function may be said to be a property of the artefact considered on its own. But we have to be careful about the meaning of the phrase "on its own" here. If we consider a technical artefact by itself *qua technical artefact*, then we have to take into account its function (otherwise we are not dealing with a technical artefact). Its function, though, is a relational property, a property that refers to human intentions.

On this dual-nature interpretation, a technical artefact considered on its own may be said to be related to particular ends, namely the ends associated with its function, and this may be a ground for attributing moral significance to a technical artefact on its own. Yet, a technical artefact has no ends of its own in the sense of intrinsic ends. The ends associated with technical artefacts are always ends of human agency and any moral significance attributed to these ends ultimately derives from the moral significance of human ends. I call this *inherent* moral significance as distinct from intrinsic moral significance (see the discussion in section 6.4). This form of moral significance is inherent because it belongs to technical artefacts by themselves, but not intrinsic because it is ultimately dependent on the moral significance of human ends.

Let us return to Latour's example of the speed bump in order to illustrate the relevance of human intentions for being a technical artefact (of a certain kind) and for the moral significance that may be attached to technical artefacts. Imagine the following situation. You are driving along a deserted Dutch country road and notice ahead a bump in the road (situation A). Nobody intentionally made this bump. It is there because of movement in the underground of the road; let's call it a 'natural' bump. You know that when you will continue driving at the present speed you will seriously damage the suspension of your car. You know this because the first time you passed this bump you ended up with a high bill from your car repair shop. So, what do you do? You slow down, take this hurdle carefully and continue your way. After a while you enter a village and approach a speed bump in the road which is placed next to a school for children (situation B). The speed bump is marked by a sign on the side of the road. By sheer coincidence, however, it has the exact same physical appearance as the bump in the road in situation A. You pass it safely with reduced speed and afterwards continue your way.

The difference between situations A and B is that in B we are dealing with a technical artefact, whereas in A we are not. The bump in the road in situation A has no function; it is not *for* anything. In situation B the bump has a kind-proper function, namely to slow down traffic in order not to endanger children. This function is partly constituted by the physical properties of bump B and partly by the intentions of the people who made the speed bump.[20] This function makes bump B into a

[20] In case by chance there was already a 'natural' bump in the road at the right place near the children school and of the right form, intentions are sufficient to turn that bump into a technical 'artefact' (no physical action or work is then required; compare the discussion about shells in chapter 4). Nevertheless, the speed bump as a technical artefact is in that case constituted by its physical structure and human intentions.

technical artefact of a certain type; it makes bump B into a speed bump. In order to underline that they are technical artefacts, signs are often put on the speed bumps themselves; these signs are there for the same reason that for instance a policeman is given a uniform, namely to make their function easily recognizable.[21]

What about the inherent moral significance of bump A and B? Bump A is simply a physical object and as such has no inherent moral significance. Bump B, on the contrary, identical to bump A as a physical object, is a technical artefact because of its intentional and material history. As a technical artefact it has a function which ties it to particular ends of human action. This implies that bump B as a technical artefact has inherent moral significance; considered by itself, as a technical artefact, it is not a morally neutral object.

Now suppose we replace bump A by B and vice versa. So we end up with a speed bump on a deserted Dutch country road and the natural bump near the school. The speed bump on the deserted road retains its inherent moral significance (the end associated with its function has not changed) but in this context its only effect will be to enrage car drivers. This effect of the speed bump may be considered to be morally problematic; this, however, is not a form of inherent moral significance, but is due to its effects in a specific *context* in which the artefact is deployed or used. In the case of the natural bump near the children school, it is also due to the context that the natural bump has moral significance, not due to the bump itself. It makes drivers slow down and this effect of the natural bump is morally significant because by chance it is located near a school (removing the natural bump would be surely morally problematic). From a moral point of view the *outcome* of the natural bump and the speed bump near the school may be considered the same: children are protected from speeding car drivers. The moral status of the natural and the speed bump is, however, different, which is due to the difference in the kind of things involved. Only the speed bump may be claimed to have an inherent moral significance. This inherent moral significance does not presuppose, however, that the technical artefact has moral agency or intrinsic ends. The origin of this significance lies in human agency and human ends.

Just as Latour, I make a 'constitutive' move in order to interpret the moral significance of technical artefacts: I take technical artefacts to be partly constituted by human intentions. It is, however, a much more modest move than his, in that it does not assume that human beings are co-constituted by technical artefacts in particular situations. Moreover, the interpretation of technical artefacts as mind-dependent, as co-constituted by human intentionality, is in a significant way different from Latour's claim that humans co-constitute technical artefacts in associations. For Latour (1999, p. 192) "a forsaken gun is a mere piece of matter"; a gun by itself, apart from the actual associations in which it occurs, is not a gun.[22] The mind-dependence of

[21] There is yet another reason for putting signs on the speed bump; in Latour's terms, part of the 'program of action' of the speed bump can be delegated to these signs, which means that the speed bumps can be made less steep (to avoid unnecessary damage to cars).

[22] This raises the question how a piece of matter, that enters an association with a human being, can co-constitute the human being as a gunman and itself as a gun.

technical artefacts on the dual-nature interpretation or the interpretation put forward by Thomasson, however, does not imply that technical artefacts exist only in situations in which human beings/intentions are co-present. Thomasson's theory of technical artefacts stresses the constitutive role of intentions in the history of these objects; the dual-nature conception of technical artefacts takes over this idea (see section 4.3). So a forsaken gun, a gun that is never used, is still a gun on account of its intentional and material history. This means that when a person uses a gun, that gun is not necessarily constituted as a gun by the intentionality of that user. So the gun may enter the human-gun system (association, collective) as a gun, and not simply as a piece of matter.[23] Moreover, as a gun it may influence the behaviour of the human involved in morally significant ways, not only because it offers options for physical action that would not be possible without it, but also because of the meaning (symbol of power) attached to the gun.

6.7 The meaning of technical artefacts

In analysing the way technical artefacts influence human behaviour, one may focus on the way in which they condition human behaviour in the context of exploiting their technical functions. This is the most direct form in which technical artefacts change human behaviour; it takes place at the level of the physical interaction between technical artefacts and human beings. It is probably one of the best researched areas of how technical artefacts condition and change human behaviour. It plays an important role in the study of man-machine interactions, which focuses mainly on the physical and cognitive skills necessary for operating machines. There is, however, yet another way in which technical artefacts influence human behaviour, namely through their *meaning*. Here the interaction is primarily of a symbolic nature.

Examples of technical artefacts with meaning attached to them abound. For many owners of a car, a car is not simply a technical artefact in the sense of a means of transport. A car may also be a status symbol or stand for a certain life style. In other words, the car has a certain meaning for its owner and this meaning may have a strong impact on the behaviour of its owner, including the way the car will be used. Take a round table used in a multi-lateral negotiation setting. It is not only a physical structure with a practical function, but it has also a meaning attached to it; the roundness of the table signifies that all the participants at the table are on an equal footing. Or, again, take a gun; it not only carries bullets to shoot, but also meanings; for some a gun signifies security, for others danger. Similar remarks apply to buildings; apart from fulfilling practical/technical functions, buildings may have meanings (think of cathedrals, high-rise buildings and so on). In all these cases,

[23] From an epistemological point of view, it may be an objective fact for the human being involved that the object is a gun. As Thomasson stresses, the mind-dependence of technical artefacts does not exclude an objective epistemology about them (see section 4.3).

the meanings of technical artefacts may a have strong influence on the way humans behave not only with regard to each other, but also with regard to the technical artefacts involved.[24] For understanding what technical artefacts do to human beings and what their moral status is, their meanings may be as important as their functions.

So, apart from the functions of technical artefacts also their meanings may be taken as a ground for claiming that technical artefacts by themselves may be assessed from a moral point of view. It may be claimed, for instance, that the meaning associated with a torture instrument makes it a morally bad instrument irrespective of its physical/technical properties (it is always possible to imagine situations in which these properties are exploited in a way that is morally good). Similarly, the meaning associated with a life-saving device (for instance, a pacemaker) makes it a morally good instrument irrespective of its specific physical and technical properties (which again in specific contexts may be exploited in morally bad ways).

The idea that technical artefacts 'have' meanings raises many philosophical issues. What kind of meaning is involved and how does it relate to, for instance, the meaning of assertions or of words or symbols in general? How do technical artefacts acquire meaning and what is the role of human intentionality therein? Is it possible for technical artefacts to have meaning by themselves? A discussion of the moral significance of technical artefacts in virtue of their meaning will require an examination of these questions. Such an examination, however, falls outside the scope of this book. I restrict myself here to a brief discussion of some points that I think are of relevance for an analysis of the moral significance of technical artefacts that is attached to their meaning. The first point is that it is not always possible to make a clear distinction between a technical function and the meaning of a technical artefact, the second that various forms of meaning have to be distinguished. The third one touches upon the role of the meaning of technical artefacts in engineering practice. My final point concerns the role of technical artefacts in observing and giving meaning to the world.

To begin with, at first sight there seems to be a clear-cut distinction between the function of a technical artefact and its meaning, because the former is intimately tied to causal interaction and the latter to symbolic interaction between the technical artefact, its users and the environment. However, technical function and meaning may become intertwined to such a degree that in particular situations it may be impossible to determine whether one is dealing with an object that performs its function through causal or symbolic interaction. Searle's distinction between causal and status functions may help to illustrate this. We have seen (section 3.5.1) that, following Searle, technical functions may be characterised as *causal* functions, that is, functions that are realized by the physical properties of the objects to which the functions are attributed. Leaving aside malfunction, a given technical function can therefore only be attributed to objects that have the right physical capacities to perform this function. This is not the case for *status* functions. A status function (e.g., being a ten Euro bill) is attributed to an object by virtue of collective intentionality

[24] For a discussion of how meanings influence human behaviour, see Verbeek (2005, chapter 7).

and can perform this function only on the basis of this collective intentionality, not on the basis of its physical capacities. Therefore, the relation between a status function of an object and its physical characteristics may be very weak.[25] A specific kind of status function is a symbolic function, where the function of the object is to express some meaning.

Searle's example of the boundary wall, discussed in section 2.1, illustrates that there may be a continuum running from causal (technical) functions to status (symbolic) functions (see (Searle 1995, p. 39–40)). The wall has the function of preventing people from crossing a boundary; it performs this function on the basis of appropriate physical properties. A boundary post or landmark has the symbolic function of marking a boundary and may well have the same function, namely preventing people from crossing the boundary. We can imagine a continuum of cases in between these two such that going from one end of the continuum to the other a technical function changes imperceptibly into a symbolic one or vice versa. In between there will be cases in which it will be difficult to determine whether the function of the artefact is a technical or symbolic one and consequently whether we are dealing with a technical artefact or a symbolic artefact.

This possibility of a continuous transition from technical functions to symbolic ones (meanings) does not imply that the distinction between words (signs/symbols) and technical artefacts cannot be made with regard to technology, as Latour (Latour 1992, p. 244) claims:

> The distinction between words and things is impossible to make for technology because it is the gradient allowing engineers to shift down – from words to things – or the shift up – from things to signs – that enables them to enforce their programs of action.

Latour rightly claims that the program of action of keeping people out of a place can be implemented through all kinds of delegation, running from human beings guarding the place at one extreme, a physical wall at the other extreme and symbolic delegations in between.[26] But that does not mean that the distinction between words and technical artefacts (and human beings for that matter) makes no sense within engineering. To begin with, the conclusion is problematic since Latour may be charged with committing a Sorite fallacy. But more importantly, not all engineering problems are of the same nature as keeping people out of a place, or regulating traffic at a dangerous spot on a road. These problems are about regulating and disciplining the behaviour of people and may be solved in ways that involve only signs.[27] It may be questioned whether such 'semiotic' solutions are engineering solutions at all, since they do not involve the design and manufacture of technical artefacts and the enforcement of the desired behaviour is based on social rules. In my opinion, the

[25] The performance of symbolic functions by physical objects may require features that relate to their physical characteristics; for a discussion of performance criteria for symbolic functions, see (Schiffer 1992, p. 134–135).

[26] See also my discussion about the various kinds of solutions to practical problems in section 2.1.

[27] Note that many of the examples Latour uses to illustrate his approach are of this nature.

making of symbols is not an engineering activity *per se*. Engineering is completely different from a semiotic enterprise, that is, an enterprise that deals with signs and words only. One cannot solve the problem of putting a man on the moon by words (signs) alone.[28] The conclusion to be drawn from the foregoing for analyses of the moral significance of technical artefacts is that in general the moral significance due to their functional/causal impact cannot be reduced to the moral significance related to their meaning or vice versa.

Our second point concerns the fact that the (symbolic) meaning of technical artefacts comes in various forms, which it may be necessary to distinguish in discussions of the moral significance of technical artefacts. According to Schiffer (1992, p. 10 ff), for instance, three different kinds of functions of artefacts may be distinguished. Next to the functions of technical artefacts as I have discussed them in this book, and which he calls technofunctions, Schiffer distinguishes sociofunctions and ideofunctions, which are related to their symbolic meanings.[29] Artefacts with technofunctions are involved in activities that change the physical environment, mainly the transport, storage and change of materials (pottery, axes etc.). Artefacts with sociofunctions occur in activities in which social positions are marked by signs or symbols, that is, in situations in which social facts are made explicit without the use of words (uniforms, insignia etc.). Artefacts with ideofunctions "encode or symbolize ideas, values, knowledge, and information" (Schiffer 1992, p. 11). Monuments, pieces of art, documents etc. are objects that serve ideofunctions. One object may perform different functions at the same time. For instance, the chair of the Pope not only physically supports the Pope (technofunction), and symbolizes his social position as head of the Catholic Church (sociofunction), but also symbolizes the position of the Pope in regard to God (ideofunction). Schiffer advocates the development of a unified behavioural science in order to study the combined effect of technofunctions, sociofunctions and ideofunctions of technical artefacts on human behaviour (Schiffer 1992, p. 141).

For studying the moral significance of technical artefacts, the distinction between socio- and ideofunctions may be relevant. Some sociofunctions as defined by Schiffer are strongly related to the efficient organization of social behaviour. These sociofunctions may often 'cross over' to technofunctions, that is, be delegated to technical artefacts. Whether this is also the case for ideofunctions is an open matter. It may be that sociofunctions and ideofunctions influence human behaviour in different ways, a fact that may be of relevance for analysing the moral significance of

---

28 See also the discussion in (Pickering 1995, p. 9–20). Hutchby (2001) rejects the idea of technology as text because it cannot explain or deal with the different affordances of technical artefacts, that is, the different possibilities they offer for action. These affordances enable and constrain the possible uses and meanings of technical artefacts.

29 Searle distinguishes between symbolic, deontic, honorific and procedural status functions that may be attributed to objects; see (Searle 1995, p. 99 ff). For a discussion of various kinds of functions of technical artefacts, including semiotic functions, from the point of view of (engineering) design, see (Muller 2001); see also (Verbeek 2005, p. 204–207). Crilly (2010) distinguishes between technical, social and aesthetic functions.

technical artefacts with socio- and ideofunctions. For the moment, however, we have to observe that a generally accepted philosophical framework for analysing the meaning of technical artefacts is lacking. Whether Schiffer's classification of symbolic meanings is a fruitful starting point for studying the moral significance of technical artefacts remains to be seen.

We come to our next point, the role of the meaning of technical artefacts in engineering practices. In many traditional 'hard core' engineering practices, such as the design of a new kind of microchip, a gas discharge lamp or an electromotor, the meaning of technical artefacts plays only a minor role, if any role at all; technofunctions instead of symbolic socio- or ideofunctions matter.[30] In others, however, their meaning may become of primary importance. Take industrial engineering or architecture. In these engineering fields, meaning may become so important that new technical artefacts, such as buildings, are taken to be 'statements', stressing that what they are about is meaning and not so much technical function. Designing technical artefacts then becomes close to being an art, with aesthetic values often taking over as the most important assessment criteria for evaluating technical artefacts. Within such contexts, engineers are very much concerned with designing meaning into their technical artefacts (see (Verbeek 2005, chapter 7). It is an interesting question to what extent that can be done at all. Independently of the meaning designed into a technical artefact, users may also attribute meaning to them. It appears that the meaning of technical artefacts is much less under control of engineers than technofunctions and the physical structures that realize those technofunctions. On the basis of a number of case studies, notably the development of the bicycle and Bakelite, Bijker (Bijker et al.1987; Bijker and Law1992), for instance, has argued that relevant social groups may play crucial roles in the interpretation and definition of technical artefacts. He has introduced the notion of interpretive flexibility, according to which different social groups may attach different meanings (functions) to the same technical artefact.

Whatever may be the role of designers or users in determining the meaning of technical artefacts, it is important to note that just as their functions, the meanings of technical artefacts appear to be relational properties. In so far as we can make sense of the notion of technical artefacts having meanings, they have these meanings only in relation to human beings; meanings are attributed to technical artefacts by human beings.[31] Consequently it makes no sense to claim that technical artefacts have moral significance on their own due to their meaning, if the phrase "on their own" is taken to mean independent of human beings. So, meaning does not lead to intrinsic moral significance, but does it lead to inherent moral significance? The answer to that

[30] What I have in mind here is meaning that is relevant for potential users. It is clear that even in hard core engineering practices a particular design or artefact may acquire a specific meaning for engineers (such as "being a masterpiece").

[31] According to Grice (1989 (1957)) physical objects or phenomena may have 'natural' meanings (e.g., smoke means fire) independent of human intentionality. I leave the possibility that similarly technical artefacts may have meanings by themselves (independent of human intentionality) out of consideration.

question depends on the role of meaning in defining technical artefact kinds. To what extent, if at all, can meaning (next to technical function and physical structure) be considered as constitutive for being a technical artefact of a certain kind? If so, our definition of technical artefact (kinds) would have to be revised to take this constitutive role of meanings into account. Take a prison. At first sight it appears that a strong case can be made for the claim that what makes a building (in the sense of a physical structure) into a prison is not just its physical structure and its function to lock up people, but also the meaning involved. If indeed in certain cases meaning may be constitutive for being a technical artefact of a certain kind, then such technical artefacts may also have inherent moral significance by virtue of their meaning. However, it may be questioned whether one is still dealing with a technical artefact proper if meaning plays a constitutive role, since then the artefact involved is a technical and symbolic artefact at the same time. In the next section I will discuss in detail an example in which various meanings have been associated with an artefact.

Finally, there is yet another way in which technical artefacts may influence human behaviour. It concerns the role of technical artefacts in observing and interpreting the world. Our perception of the world is to a large extent mediated by technical artefacts, and this mediating role of technical artefacts may influence the way we give meaning to the world (Ihde 1990; Verbeek 2005, Ch. 4). Verbeek (2008) argues that obstetrical ultrasound technology, for instance, does not simply make visible the unborn child in the womb, but that this technology also shapes how the unborn child is perceived; he claims that the technology constitutes the foetus and the parents in specific ways. In these cases it is not the meaning of the technical artefact itself that is at stake, but the interpretation of objects and phenomena observed with the help of it. If technical artefacts indeed influence the way meaning is attributed to the world, then in my opinion this may be a way they shape human behaviour that, though indirect, can nevertheless be even more far-reaching than the ways discussed above.

6.8 An example: the meaning of buildings

The following example is intended to illustrate that meanings attributed to artefacts may play an important role in how we conceptualize the artefacts we are dealing with. The example concerns a public-housing project, called Cité de la Muette, at Drancy, not far from Paris, during the 1930's.[32] The use and meaning of this public-housing estate changed in a dramatic way, when its social housing function was replaced by another function, namely that of a concentration and transit camp for the deportation of Jews during the Second World War. Long after this war it acquired again a new meaning as a national monument.

[32] For more historic details about this example, see (Priemus and Kroes 2008). The analysis presented here is different from the one in that paper.

Cité de la Muette is one of the French *grands ensembles*, developed in the mid 1930s. It comprised some 1250 apartments, mostly in three or four-storey U-shaped buildings, but it also included five high-rise towers, each of fifteen storey's. In contemporary professional journals as the *The Architectural Forum* and *The Architects Journal*, the estate is described as modern and innovative. Its construction relied largely on prefabricated concrete elements, this being a labour-saving approach, which would result in modern amenities. Cité de la Muette was considered as a promising, revolutionary housing-estate which would give an impetus to French modern housing in general. Rowse (1934, p. 201) gives the following passionate description of this public housing project:

> The drama of the five great towers, whether seen from the air, or the ground, standing like sentinels amidst confusion, cannot be denied. Perhaps they point to a road we all may follow to a sanity, order and beauty in urban design which may prove the salvation of future generation of town dwellers who otherwise would sink, crushed beneath the soul-destroying banality of the 'Housing Estate'.

Architects as well as engineers considered Cité de la Muette a symbol of progress in public housing.

Not so its tenants and occupants, as it turned out. The economic climate in the 1930s was bad, to say the least. Unemployment rose rapidly and the rent and service costs in Cité de la Muette were far higher than in other public-housing estates. As many of the amenities that were planned for the residential environment had never materialized, the residents of Cité de la Muette perceived the price-quality ratio as far lower than the architects and the engineers. A huge gap had emerged between the meaning in the eyes of the architects and engineers on the one hand (roughly, modern, progressive social housing) and in the eyes of the occupants (and home seekers in the area) on the other (expensive, isolated housing). This was reflected in a chronic high non-occupancy rate that in itself had an adverse effect on the living climate. Consequently, for the *Public Housing Authority of the Department of the Seine*, the housing estate of Cité de la Muette was something of a failure as a social housing project, due to the fact that many of the houses could not be rented once the project was completed and remained empty thereafter, despite the generous rent subsidies available.

A second episode in the life of Cité de la Muette started with the onset of the Second World War. In 1939, before the Nazis invaded France, the French government opened camps such as those at Gurs and Noe, designed to accommodate Spanish refugees fleeing from the Franco regime. Cité de la Muette, also known as 'Drancy', became one of those camps guarded by the French police. The German army took control of the camp in June 1940 and made it into a concentration camp for Jewish prisoners, their last station before being transported to Germany. In a period of three years (August 1941 to August 1944) some seventy thousand people were detained here. Of all French Jews, transported to Auschwitz, 83% came from Drancy. The first transport carried one thousand Jews to Auschwitz on 22 June 1942. The final transport took place on 31 July 1944. It is in this period that Drancy became a symbol of terror and oppression.

What happened after the war is that at least up until the beginning of the twenty-first century, the U-shaped rectangle of Cité de la Muette was still in use as a housing facility, recently for marginalized people waiting for better housing (Bourgon 2002). The five towers were used by the gendarmerie after WWII, then acquired by the Ministry of Défence, and finally demolished in 1976. According to Bourgon, the re-instalment as a social housing facility took place in the context of the housing crisis just after the war. She also remarks that in May 2001 the French government officially declared what remained of Cité de la Muette a historic monument, not so much because of its role in the Second World War, but because it is considered as a symbol of social housing in the twentieth century.

Here we encounter a case in which buildings have acquired different functions and meanings in the course of their life. How is the moral significance of these buildings to be interpreted, especially in the face of the morally conflicting meanings attached to them? That is a rather complex issue. For a start, it is important to note that the buildings underwent this change in use and meaning without any substantial changes to their physical structure. It is no exception that houses and other buildings, for some time or permanently, undergo a change in function and meaning: dwellings may become shops, churches may become apartment buildings. If these changes in function are the result of a thorough re-design and restructuring of the original buildings, then this is not a noteworthy feat. Indeed, designing is the act of creating objects that may perform a certain function, and so it comes as no surprise that re-designing a building may result in the creation of a new function and corresponding meaning. But as this case shows, such changes in function and meaning may also occur without significant re-designing and restructuring of the original building.

If we assume that throughout the whole history of Cité de la Muette we are dealing with the same physical buildings, how can we understand that they can serve functions that are morally so conflicting? The obvious answer, it seems, is to appeal to the neutrality-thesis of technical artefacts and to claim that the buildings themselves are morally neutral. It is the use humans make of them that can be qualified in a moral sense as good or bad. However, this answer is somehow unsatisfactory. If these buildings are taken to be a public-housing estate then they inherently have morally good intentions associated with their kind-proper function, but when they are taken to be a concentration camp the opposite is the case. So it seems difficult to maintain that these buildings, when they are taken to be physical constructions with a kind-proper function, that is, when they are taken to be technical artefacts, are morally neutral. As such, they have inherent moral significance. The neutrality theses applies to the buildings of Cité de la Muette only in so far as these buildings are taken to be physical constructions and nothing more, that is, as objects without any kind-proper function.

So the question about how to interpret the moral significance of the buildings of Cité de la Muette hinges on how the notion of buildings is to be interpreted. The crucial question is what kind of object these buildings are taken to be. If we consider them to be physical objects, they are morally neutral. If we take these buildings to be *technical artefacts* and if we apply our theory of technical artefact kinds, then these buildings have been, throughout their whole life, an instance of the technical artefact kind 'social-housing estate' since social housing was the kind-proper function that the designers and makers had in mind when realizing the estate. As such they may be taken to have an inherent positive moral significance. From this point of view, the buildings were used in a morally bad way *as* a concentration camp during the Second World War, which use involves the assignment of a use-proper or use-accidental function.

What may be considered problematic about taking the buildings to be technical artefacts is that this interpretation does not take into account the meanings associated with the buildings. Houses and buildings are more than just technical artefacts, than physical constructions performing technical functions. The function of a house involves much more than only technical aspects – protection against rain, wind, cold et cetera; it also involves psychological and social characteristics – providing security and safety, signalling status et cetera. A clear separation between these technical and psychological and social aspects is often hard to make; security and safety are clearly related to technical characteristics of a house, but also to how a house is experienced and given meaning by its occupants. As I pointed out earlier, there is no sharp dividing line between functions and meaning.

Now suppose that the meaning associated with an object may be, together with its physical structure and technical function, constitutive for the kind of object we are dealing with.[33] After all, the meaning of an object may significantly influence the way we deal with it. Cité de la Muette provides a clear illustration of this. During the war it had become a symbol of oppression. It is this symbolic meaning that has played an important role in post-war discussions about what to do with Cité de la Muette. On the interpretation of the buildings as mere technical artefacts this is hard to understand. Why would it be a problem to return to the 'proper' use of the buildings as a social-housing estate intended by its designers and makers? Apparently, Cité de la Muette was more than simply a technical artefact. The dramatic change in use and meaning during the wartime had made the buildings into something else, namely into a concentration camp – it was not something used *as* a concentration camp. This suggests an alternative to the above interpretation. Instead of being technical artefacts, the buildings may be taken to be artefacts of a different kind, namely artefacts that are co-constituted by their physical construction, their function and their meaning. For lack of a better name this kind of artefacts may be referred to as 'techno-symbolic' artefacts. Thinking along these lines, schematically at least three different techno-symbolic artefacts may be taken to figure in the story about Cité de la Muette, namely a social-housing estate, a concentration camp and a historic monument (see Fig. 6.1). Each of these techno-symbolic artefacts may be claimed to have inherent moral significance.

Clearly, the idea of techno-symbolic artefacts will have to be worked out in more detail, in particular the roles that physical structure, technical function and meaning play in constituting something as a techno-symbolic artefact. Here I will restrict

[33] As the example of Duchamp's *pissoir* shows, meaning may be the prime factor determining the kind to which an object belongs; Duchamp did not use a technical artefact *as* a piece of art, but made it into a piece of art by giving it that status or that meaning.

	Cité de la Muette 1 Social Housing Estate	Cité de la Muette 2 Concentration Camp	Cité de la Muette 3 Historic Monument
Physical structure	U-shape building, four stories high etc.	*Idem*	*Idem*
'Technical' function	Housing people	Imprison people	(Housing people)
Meaning	Symbol of Modernity	Symbol of terror	Symbol of progress in social housing

Fig. 6.1 Cité de la Muette as three different kinds of techno-symbolic artefacts

myself to the following two brief remarks. First, just like technical artefacts, techno-symbolic artefacts have a dual nature. From the dual-nature perspective the distinction between function and meaning is not significant; both imply reference to human intentions. Second, our theory of technical artefact kinds cannot be projected simply onto techno-symbolic artefacts. The reason is that changes in use and meaning of physical constructions may lead to instances of new kinds of techno-symbolic artefacts, whereas changes in use do not lead to instances of new technical artefact kinds. This difference is connected to the fact that the relation between a meaning and a physical construction is different in nature from the relation between a technical function and a physical structure. A technical function is in Searle's terminology a causal function, which means that it is (to be) realized by its physical structure, whereas meanings relate to status functions and (collective) intentionality. Because of this, the link between meaning and physical structure is much weaker than the link between technical function and physical structure. This means that the kind of activities necessary for creating new (instances of) techno-symbolic artefact kinds may be different from the activities necessary for creating new (instances of) technical artefact kinds.

6.9 Conclusion

The main conclusion to be drawn is that any attempt to analyse the moral significance of technical artefacts considered as objects on their own, in the sense of objects separated from human intentionality (agency), is on the wrong track. The reason is that such a notion of technical artefacts, as objects on their own, makes no sense. I have argued that the notions of technical artefacts underlying the positions that technical artefacts are morally neutral objects or objects with intrinsic moral significance are highly problematic. I have also reviewed Latour's proposal to treat humans and technical artefacts symmetrically, which implies attributing moral agency to technical artefacts. In my opinion this proposal does not bring us any closer to a better understanding of the moral role of technical artefacts. Next, I have argued that if we take account of the constitutive nature of human intentions for technical artefacts, they may be said to have an inherent moral significance that is not intrinsic, because it finds its origin in the moral significance of human agency. I have also explored the importance of the meaning of technical artefacts for understanding their moral significance. My overall conclusion is that there is no need to attribute to technical artefacts intrinsic forms of agency or intrinsic ends, that is, forms of agency or ends that are isolated from human agency and human intentions.[34] Nevertheless, it is possible to maintain that technical artefacts have inherent moral significance.[35]

Let me end this chapter by positioning the problem of the (moral) agency of technical artefacts in a somewhat broader perspective. This issue is part of a more general problem related to the outcomes of human work and action. At the origin of this problem lies the observation that the outcomes of human action tend to 'live a life of their own'. This observation not only applies to action that results in material objects such as technical artefacts (what Aristotle called productive action or 'poesis'), but also to action that brings forth immaterial results (non-productive action, 'praxis'), such as a promise, or the election of a president. Technical artefacts, once set in the world, may cause effects never intended by their designers, makers or users. The same applies to the results of non-productive acts: the outcome of one act may lead to another act, either performed by the same person or another, which again may lead to another and so on, resulting in actions or state of affairs never intended by the original agent or any of the other agents involved. The outcomes of human action, whether material or immaterial, appear to take on a form of agency of their own, which is beyond the control of human beings. This is a feature of what Hannah Arendt (1958) calls the Human Condition. The artificial world, the world of human making and doing, somehow in part escapes human control. For some that may be frustrating and hard to accept, but I do not see what is to be gained by attributing agency to the things we make, in particular moral agency. On the contrary, when we extend moral assessment to technical artefacts on account of their agency,

[34] This conclusion runs parallel to the conclusion we have drawn with regard to the normativity of technical artefacts; see section 4.1.

[35] Johnson (2006) comes to more or less the same conclusion but on the basis of a different analysis. She claims that technical artefacts, computers in particular, are moral entities without being moral agents. They are moral entities because they have an inbuilt intentionality which is related to their function. Because of this intentionality, technical artefacts, as opposed to natural objects, belong to the realm of morality. They are not, however, moral agents because they lack the capacity of intending to act.

we may end up in situations in which moral shortcomings (in the acts) of technical artefacts are considered to be the cause of harm done to people. Technical artefacts, however, cannot be held morally accountable for any damage. The attribution of moral agency to technical artefacts may make us blind for the responsibility we bear for the world of our own making. As Johnson (2006, p 204) states:

> I believe that attributing independent moral agency to computers is dangerous because it disconnects computer behavior from human behavior, the human behavior that creates and deploys the computer systems. This disconnection tends to reinforce the presumption of technological determinism, that is, it reinforces the idea that technology has a natural or logical order of development of its own and is not in the control of humans.

This not only applies to computers, but to technical artefacts in general. Contrary to what Johnson suggests, I do not think that humans are in control of technology (at least not fully). Be that as it may, I am in favour of an "ethical heuristics" that strives for an approach to the moral status of technical artefacts in which humans bear as much as is reasonably possible the moral burden of what we do to each other, with or without the help of technical artefacts.

References

Achterhuis, Hans. 1995. De moralisering van de apparaten. *Socialisme en Democratie* 52 (1):3–12.

Akrich, Madeleine. 1992. The de-scription of technical objects. In *Shaping technology/building society; studies in sociotechnical change*, eds. Wiebe E. Bijker, and John Law, 205–224. Cambridge (Mass.): MIT Press.

Arendt, Hannah. 1958. *The human condition*. Chicago: The University of Chicago Press.

Bijker, W.E., T.P. Hughes, and T. Pinch. 1987. The social construction of technological systems: new directions in the sociology and history of technology. Cambridge (Mass.): The MIT Press.

Bijker, W.E., and J. Law. 1992. Shaping technology/building society: studies in sociotechnical change. Cambridge (Mass.): The MIT Press.

Bourgon, A. 2002. La Cité de la Muette à Drancy: ambiguïtés, difficultés et perspectives de l'heritage. In *Bulletin no. 50–51*.

Butler, Samuel. 1985 (1872). *Erewhon*. London: Penguin Books.

Crilly, Nathan. 2010. The roles that artefacts play: technical, social and aesthetic functions. *Design Studies* 31 (4):311–344.

Ellul, Jacques. 1964. *The technological society (trans. John Wilkinson)*. New York: Alfred A. knopf.

Floridi, Luciano, and J.W. Sanders. 2004. On the morality of artificial agents. *Minds and Machines* 14 (3):349–379.

Grice, P. 1989 (1957). Meaning. In *Studies in the way of words*, ed. P Grice. Cambridge (Mass.): Harvard University.

Hutchby, Ian. 2001. Technologies, texts and affordances. *Sociology* 35 (2):441–456.

Huyke, Héctor José. 2003. Technologies and the devaluation of what is near. *Techné* 6 (3):57–70.

Ihde, Don. 1990. *Technology and the life world*. Bloomington/Minneapolis: Indiana University Press.

Introna, Lucas D. forthcoming. Towards a post-human intra-actional account of socio-technical agency (and morality). In *Technical artefacts and moral agency*, eds. Peter Kroes, and Peter-Paul Verbeek. Springer.

Johnson, Deborah G. 2006. Computer systems: moral entities but not moral agents. *Ethics and information technology* 8:195–204.
Latour, Bruno. 1992. Where are the missing masses? The sociology of a few mundane artifacts. In *Shaping technology/building society; studies in sociotechnical change*, eds. Wiebe E. Bijker, and John Law, 225–258. Cambridge (Mass.): MIT Press.
Latour, Bruno. 1993. *We have never been modern*. Trans. Catherine Porter. New York: Harvester Wheatsheaf.
Latour, Bruno. 1999. *Pandora's hope; essays on the reality of science studies*. Cambridge (Mass.): Harvard University Press.
Latour, Bruno. 2002. Morality and technology; the end of the means. *Theory, Culture & Society* 19 (5/6):247–260.
Latour, Bruno. 2005. *Reassembling the social; an introduction to actor-network-theory*. Oxford: Oxford University Press.
Muller, W. 2001. *Order and meaning in design*. Utrecht: Lemma.
Pickering, Andrew. 1995. *The mangle of practice; time, agency & science*. Chicago: The University of Chicago Press.
Pitt, Joseph C. 2000. *Thinking about technology; foundations for the philosophy of technology*. New York: Seven Bridges Press.
Pitt, Joseph C. forthcoming. "Guns don't kill, people kill"; values in and/or around technology. In *Technical artefacts and moral agency*, eds. Peter Kroes, and Peter-Paul Verbeek. Springer.
Priemus, Hugo, and Peter Kroes. 2008. Technical artefacts as physical and social constructions: the case of Cité de la Muette. *Housing Studies* 23 (5):717–736.
Rowse, E.A.A. 1934. Housing in Paris, the Cité de la Muette Drancy, Housing Supplement no. 5. *The architects journal* 80:195–201.
Schapiro, Tamar. 2001. Three conceptions of action in moral theory. *Noûs* 35 (1):93–117.
Schiffer, Michael Brian. 1992. *Technological perspectives on behavioral change*. Tucson: The University of Arizona Press.
Searle, John. 1995. *The Construction of Social Reality*. London: Penguin Books.
Verbeek, Peter-Paul. 2005. *What things do; philosophical reflections on technology, agency, and design*. Pennsylvania: The Pennsylvania State University Press.
Verbeek, Peter-Paul. 2008. Morality in design; design ethics and the morality of technological artifacts. In *Philosophy and design: from engineering to architecture*, eds. Pieter E. Vermaas, Peter Kroes, Andrew Light, and Steven A. Moore. Dordrecht: Springer.
Von Wright, Georg Henrik. 1963a. Practical inference. *The philosophical review* 72 (2):159–179.
Von Wright, Georg Henrik. 1963b. *The varieties of goodness*. London: Routledge & Kegan Paul.
Wiegel, Vincent. 2007. Sopholab. Experimental computational philosophy. Delft University of Technology; PhD Thesis, Delft, The Netherlands.
Winner, Langdon. 1985. Do artifacts have politics? In *The social shaping of technology*, eds. D. Mackenzie, and J. Wajcman. Buckingham: Open University Press.
Winner, Langdon. 1986. *The whale and the reactor; a search for limits in an age of high technology*. Chicago: Chicago University Press.
Winner, Langdon. 1992 (1977). *Autonomous technology; technics-out-of-control as a theme in political thought*. Cambridge (Mass.): MIT Press.

Chapter 7
Epilogue

In this epilogue I start with a recapitulation of the dual-nature thesis of technical artefacts while highlighting its main innovative features in comparison to existing theories of technical artefacts and technical functions. Then, looking ahead to interesting topics of further research, I propose to shift attention from technical artefacts taken in isolation, as I have done in this book, to technical artefacts as embedded within social systems. These broader systems are usually referred to as 'socio-technical' systems. The notion of a socio-technical system opens up new and interesting ways to study the mutual interaction between the technical and the social world and is rapidly gaining currency in various fields of study. However, a clear conceptualization of this kind of system is still lacking. As their name already indicates, socio-technical systems seem to have a hybrid nature as well, but a brief look suffices to show that the kind of hybridity encountered here is different from the hybridity of technical artefacts.

7.1 The dual nature of technical artefacts

The outcome of the comparison of technical artefacts to physical (natural) and social objects is that technical artefacts are a kind of object *sui generis*. On the basis of an analysis of how technical artefacts are conceived in engineering practice, I have characterized them as human-made physical constructions with a 'for-ness' or practical function. Another way of expressing more or less the same thing is by characterizing technical artefacts as physical constructions that embody a purposeful design. As such they are neither physical objects nor social objects. A physical object does not embody or is not the realization of a purposeful design, and physical constructions play no significant role in the realization of functions of social objects. As objects with physical and functional features technical artefacts are hybrid objects. They are objects that combine two different kinds of features, each of which plays a constitutive role in being a technical artefact. The physical features of a

P. Kroes, *Technical Artefacts: Creations of Mind and Matter: A Philosophy of Engineering Design*, Philosophy of Engineering and Technology 6,
DOI 10.1007/978-94-007-3940-6_7, © Springer Science+Business Media B.V. 2012

technical artefact that realize its function are intrinsic, mind-independent properties. By contrast, its functional features are mind-dependent; they are related to human intentions, since only in relation to human intentions (purposes) physical objects may have functions. At the same time these functional features are also related to the physical features of technical artefacts because of the realization relation. So the functional features of a technical artefact act as a kind of bridge that connects its intrinsic physical features to its relational intentional features. Because of its physical and intentional features a technical artefact has a dual nature; it is truly a creation of mind and matter.

Elaborating on Thomasson's theory of artefact kinds, which covers artefacts in general, I have proposed a theory of technical artefact kinds that accounts for the dual nature of technical artefacts. To this theory I have added a theory of technical functions, since being an instance of a technical artefact kind involves the assignment of a (kind-proper) function. It is the combination of this theory of technical functions with a theory of technical artefact kinds that makes the approach chosen here different from most analyses of technical functions discussed in the literature. These analyses focus primarily on explicating the notion of an object having a technical function and typically assume that if an object has or may be ascribed a function, then that object is an instance of the corresponding technical kind. That, however, is a problematic assumption that brings the distinction between accidental and proper functions into play. In explicating this distinction a second difference between most approaches toward technical functions taken in the literature and the approach chosen here comes to the fore. Instead of explicating the notion of proper function in terms of function assignments by users, I relate proper functions to function assignment by the creators of technical artefacts. In this way I am able to solve a problem on which various theories of technical functions run afoul, namely the problem of how a technical artefact that loses its ability to perform its function (a malfunctioning technical artefact) may still remain an instance of its technical artefact kind.

One of my reasons for giving priority to the creators of technical artefacts, and not to its users, when it comes to determining what I have called the kind-proper function of a technical artefact is that the sort and amount of intellectual and physical work that goes into designing and making technical artefacts is typically in no way comparable to the work that goes into using these technical artefacts. In line with Thomasson's theory of technical artefact kind I am of the opinion that it is the creator of a technical artefact who ultimately determines what kind of new object is put into the world, not the user. Creating instead of using appears to be the kind of activity that may enrich the ontology of the world and that is exactly what engineers are doing when they design and make new technical artefacts. A closer look at the engineering practice of creating technical artefacts reveals that the creation process somehow reflects the dual nature of the objects created. Designing may be described as a process in which an object characterized only in terms of functional features is 'translated' in an object that is characterized only in terms of physical features. So, the design process moves from the domain of human intentions (purposes) to the domain of physical features (physical capacities), that is, moves between the two kinds of features that are constitutive for an object being a technical artefact. I have also argued that the idea that technical artefacts are physical objects that embody a design may be interpreted in a way that comes close to the dual nature account of technical artefacts, although it has to be admitted that the notion of a design remains rather obscure.

Any discussion of the nature of technical artefacts will have to address issues about their normativity and I have done so extensively. The principle issue here is whether technical artefacts by themselves may be the bearers of some form of goodness, and if so, what form(s) of goodness. The most obvious forms of goodness to be associated with technical artefacts are instrumental goodness and moral goodness. Discussions about how instrumental goodness of technical artefacts is to be interpreted play an important role in theories of technical functions. Independently of these discussions there is a long-standing discussion going on about the moral goodness of technical artefacts. In my opinion both discussions suffer from a common problem, namely, a lack of a clear understanding of what is meant by the expression "technical artefacts by themselves". On the dual-nature account such a notion does not make sense if 'by themselves' means 'independent of human beings'. I have argued that a technical artefact, *qua* technical artefact, may be the bearer of instrumental as well as moral goodness. Both forms of goodness, however, are not intrinsic to technical artefacts but are related to human ends. As regards the moral status of technical artefacts, one of the most interesting topics for future research is in my opinion the moral significance of the meaning of technical artefacts; just as the functions of technical artefacts also their meanings may provide reasons for how to act with regard to those artefacts.

7.2 Socio-technical systems: more hybridity

In my attempt to understand the nature of the technical world in which we live, I have focused on the question of what kind of objects technical artefacts are, assuming that these technical artefacts are the elementary building blocks of this technical world. I have analysed technical artefacts more or less as objects on their own, as isolated objects not embedded in larger technical or social systems and concluded that from that perspective technical artefacts are hybrid objects different from physical and from social objects. In the technical world the physical and social world so to speak meet. The physical world provides the physical capacities in which the efficacy of the technical world is grounded, whereas the social/intentional world furnishes the designs or substantive ideas that inextricably tie technical artefacts to human ends. Together these physical capacities and designs determine what kind of technical objects populate the technical world. This may be interpreted as implying a rather voluntaristic picture of the technical world: subject to natural and technical constraints the technical world is populated with technical artefacts that contribute to the realization of ends that humans think worthwhile to pursue (see Fig. 5.1). The technical world is therefore a world of human making and according to this line of thinking humans get, given what is physically and technologically possible, the technology they desire (and, some may be inclined to add, they thus deserve).[1]

In my opinion there is more than just a kernel of truth to the conclusion that viewed from the perspective of technical artefacts the technical world is a world of our own making. None of the technical artefacts making up the technical world is there of its own account or by its own doing. By definition they are creations of humans and reflect human needs, desires and ends. However, this story about the technical world tells at most half of the story there is to be told about the relation between humans and the technical and social world. At least it has to be complemented by another story that highlights that, as we have already observed in chapter 6, the technological world somehow appears to take on a life of its own irrespective of human needs, desires and ends or of its social context. That story may be told by shifting the focus of attention from technical artefacts on their own to the technical and social systems in which technical artefacts are embedded and to the dynamics of these technological and social systems in relation to each other. After all, technical artefacts on their own do nothing; if they are not used they do not perform their function and are socially as inert as technical artefacts on a shelf in a shop or stowed away in a storehouse. In this other story, therefore, the actual use of technical artefacts will play a much more prominent role than it has in the analysis presented so far.

In order to actually perform their function, technical artefacts not only require users, but very often also other (systems of) technical artefacts have to be in place. For instance, for me to use my computer a well-functioning electric energy supply system has to be in place. That not only requires all kind of well-functioning technical systems (energy plants, infrastructure for transporting electric energy etc.) but also well-functioning social systems (billing systems, laws and regulations with regard to energy production companies and electric energy trade, standardization systems etc.). The electric-energy supply system is an example of a socio-technical system. As we have already seen in the discussion of the various kinds of solutions that may be implemented for solving practical problems (see section 2.1), a defining feature of socio-technical systems is that their functioning depends on the functioning of social and technical systems. In my opinion an analysis of socio-technical systems and how they develop may provide the right kind of story to complement the above voluntaristic picture of the technical world. A shift of focus from technical artefacts to socio-technical systems opens up a way of studying the relation between the technical and the social world in which all kinds of social factors alongside humans needs, desires and ends may play a role in how the technical world is shaped and develops.

Precisely because the notion of socio-technical systems offers such an interesting starting point for studying the interaction between technology and the social world, this notion has become popular in recent decades among historians of technology, STS-scholars and philosophers of technology. But it is also gaining in popularity

[1] Since there is no *volonté generale* of humankind about what ends to realize by what kind of technical means it may rightly be questioned whether such sweeping statements make sense; the point I want to make here is that technical artefacts originate in the needs, desires and ends of people without going into any detail about whose needs, desires and ends and what happens in situations of diverging needs, desires and ends.

among systems engineers. Since many technical systems can perform their function only in combination with the appropriate social infrastructure (appropriate social institutions, laws etc.) the design and management of these technical systems requires that (changes in) social infrastructures are taken into account. From a systems-engineering point of view there are schematically two options for dealing with these social infrastructures. One option, which is more or less the traditional option, is to consider the social infrastructure not to be part of the system that is the object of systems engineering. In that case, system engineering focuses on the design and management of technical systems that perform their function on condition that a particular social infrastructure is in place. In this way the (required) social infrastructure is treated as a set of constraints on the technical system to be designed and managed. The other option is to make the social infrastructure part of the system under consideration (see also section 5.8). Then systems engineering deals with the design and management of socio-technical systems and will have to address not only the design of technical hardware but also the design of the appropriate social infrastructure (often referred to as 'institutional design"). Such an approach may have as an advantage that technical and social systems may be designed in a way that they are better attuned to each other than when they are designed separately. However, there are surely also practical drawbacks to the socio-technical approach given the vast differences in the way technical and institutional design have been organized and practised traditionally.

In spite of its apparent attractiveness and frequent use in these various fields, the notion of a socio-technical system raises intriguing philosophical issues. Intuitively it may seem clear what kind of system is intended. Hughes, for instance, describes them as follows (he calls them 'technological systems') (1987, p. 51):

> Technological systems contain messy, complex, problem-solving components. They are both socially constructed and society shaping. Among the components in technological systems are physical artifacts, such as turbogenerators [...] organizations, such as manufacturing firms [...] components usually labeled scientific, such as books [...] and research programs. Legislative artifacts, such as regulatory laws [and] natural resources, such as coal mines [...] An artifact – either physical or nonphysical – functioning as a component in a system interacts within other artifacts, all of which contribute directly or through other components to the common system goal.

For Hughes a socio-technical system is made up of very heterogeneous elements all of which are bound together by the overall function of the system.[2] Thus, in line with our discussion of various kinds of solutions to practical problems (see section 2.1) a socio-technical system is intuitively a means to solve a practical problem that combines technical and social elements.

---

[2] In a similar heterogeneous vein the notion of socio-technical system is defined by Geels (2005, p. viii). Also the notion of system employed in systems engineering is a heterogeneous one; according to the systems engineering handbook of the *International Council on Systems Engineering* a system is (INCOSE 2004, p. 11) "An integrated set of elements that accomplish a defined objective. These elements include products (hardware, software, firmware), processes, people, information, techniques, facilities, services, and other support elements." This INCOSE conception of a system may be not as wide ranging as Hughes's, the inclusion of people makes a system also heterogeneous.

As soon, however, as an attempt is made to make this intuitive notion of socio-technical system more precise, problems arise. For instance, there is the problem that the notion of socio-technical systems inherits from systems theory in general, namely how to draw the boundary around a socio-technical system (Kroes et al. 2006). What is considered to be part of the system and what not? It is highly questionable whether an appeal to "the common system goal", or to the function of a socio-technical system may solve this problem. In the case of a socio-technical system it is very often not clear who is the system owner or system user. Various stakeholders may be involved each of whom may have a different view on the function or system goal. As a result, these various stakeholders will draw the system boundaries in different way which means that there is no shared, unambiguous definition of the socio-technical system involved.

Even if we assume that in a particular case the function of a socio-technical system and its boundaries may be defined in an unambiguous way another conceptual problem has to be faced. This problem has to do with the heterogeneity of the elements of the system. These various kinds of elements are supposed to be bound together into one socio-technical system by its overall function or goal. By interacting with each other they contribute, according to Hughes, to the realization of the system goal. How is that to be understood? How is the interaction between these various kinds of elements to be conceptualized? In order to see the problems involved in answering these questions we have to take a brief look at the various kinds of elements that are part of socio-technical systems.

Socio-technical systems appear to be hybrid systems made up of technical and social elements. On closer inspection they turn out to be even more than hybrid systems, for they are made up of more than two different kinds of elements. If we take our lead from the description of a socio-technical system by Hughes, such a system may contain the following kinds of elements:

- Technical elements (from component level to technical systems)
- Human beings (in various roles: operators, technician, legislator,…)
- Social elements (organizations, laws, regulations,…)
- Scientific elements (books, research program,…) and
- Natural resources.

Interactions between these various kinds of elements are supposed to glue these elements together into one system. From a systems-theoretic point of view they are the relations between the elements of the socio-technical system. This implies that there are not only interactions (relations) between elements of the same kind, but also 'hybrid' interactions, that is, interactions between elements of different kinds. If not, the whole system would fall apart into a number of independent systems containing only elements of the same kind.

Leaving aside problems about how to conceptualize the various kinds of elements involved in socio-technical systems, it is far from clear what kind of interactions are keeping the system together, whether interactions between elements of the same kind are concerned (for instance, between an organization and a law) or between elements of different kinds (for instance, between a law and a technical system).

It is, for instance, an open issue whether all these various kinds of interactions, hybrid or not, can be reduced to combinations of a set of 'basic' interactions (such as physical, intentional and functional interactions). If we want to go beyond the intuitive idea of a socio-technical system in order to make it a more rigorous tool for understanding the dynamics between the technical and the social word, a conceptual clarification of the kinds of elements making up a socio-technical system and the kinds of relations between these elements is needed.[3]

In conclusion, what makes socio-technical systems such interesting systems are the hybrid relations between its technical and social elements. These relations are the channels through which the technical and social world influence each other. In order to better understand the dynamics between the technical and the social world, these hybrid relations have to be better understood, as well as the nature of the elements involved. I have concentrated on a clarification of just one of these elements, namely technical elements. Much work remains to be done with regard to the notion of a socio-technical system. It is one thing to stipulate that technical and social elements interact in socio-technical systems, it is another to come up with a clear conceptualization of this hybrid interaction. A clarification of this kind of hybridity, which is clearly different from the hybrid nature of technical artefacts, poses another exciting challenge for the for the philosophy of technology.

References

Bauer, Johannes M., and Paulien M. Herder. 2009. Designing socio-technical systems. In *Philosophy of technology and engineering sciences*, ed. Anthonie Meijers, 601–630. Elsevier.

Franssen, Maarten, and Bjorn Jespersen. 2009. From nutcracking to assisted driving: stratified instrumental systems and the modeling of complexity. Paper presented at the Second international symposium on engineering systems, MIT, Cambridge, Massachusetts, June 15–17 2009

Geels, Frank W. 2005. *Technological transistions and systems innovations; a co-evolutionary and socio-technical analysis*. Cheltenham: Edward Elgar Publ. Ltd.

Hughes, Thomas P. 1987. The evolution of large technological systems. In *The social construction of technological systems. New directions in the sociology and history of technology*, eds. W.E. Bijker, T.P. Hughes, and T.J. Pinch, 51–82. Cambridge (Mass.): MIT Press.

INCOSE. 2004. *Systems engineering handbook, Version 2a*. International Council on Systems Engineering (INCOSE).

Kroes, Peter, Maarten Franssen, Ibo van de Poel, and Maarten Ottens. 2006. Treating socio-technical systems as engineering systems: some conceptual problems. *Systems research and behavioral science* 23:803–814.

Ropohl, Günther. 1999. Philosophy of socio-technical systems. *Society for philosophy and technology* 4 (3):59–71.

Vermaas, Pieter E., Peter Kroes, Ibo van de Poel, Maarten Franssen, and Wybo Houkes. 2011. *A philosophy of technology; from technical artefacts to sociotechnical systems*. Synthesis lectures on engineers, technology and society: Morgan & Claypool.

[3] For philosophical discussions about conceptual issues with regard to socio-technical systems, see (Franssen and Jespersen 2009; Ropohl 1999; Kroes et al. 2006; Vermaas et al. 2011, Ch. 5); see also (Bauer and Herder 2009).

Name Index

P. Kroes, *Technical Artefacts: Creations of Mind and Matter: A Philosophy of Engineering Design*, Philosophy of Engineering and Technology 6,
DOI 10.1007/978-94-007-3940-6, © Springer Science+Business Media B.V. 2012

MIX
Papier aus verantwortungsvollen Quellen
Paper from responsible sources
FSC® C105338

If you have any concerns about our products,
you can contact us on
ProductSafety@springernature.com

In case Publisher is established outside the EU,
the EU authorized representative is:
Springer Nature Customer Service Center GmbH
Europaplatz 3, 69115 Heidelberg, Germany

Printed by Libri Plureos GmbH
in Hamburg, Germany